I0056254

Recent Trends in Nanobiotechnology

Food and Biomedical Applications

Recent Trends in Nanobiotechnology

Food and Biomedical Applications

Prakash Saudagar
K. Divakar

CWP
Central West Publishing

This edition has been published by Central West Publishing, Australia
© 2019 Central West Publishing

All rights reserved. No part of this volume may be reproduced, copied, stored, or transmitted, in any form or by any means, electronic, photocopying, recording, or otherwise. Permission requests for reuse can be sent to editor@centralwestpublishing.com

For more information about the books published by Central West Publishing, please visit https://centralwestpublishing.com

Disclaimer
Every effort has been made by the publisher, editor and authors while preparing this book, however, no warranties are made regarding the accuracy and completeness of the content. The publisher, editor and authors disclaim without any limitation all warranties as well as any implied warranties about sales, along with fitness of the content for a particular purpose. Citation of any website and other information sources does not mean any endorsement from the publisher and authors. For ascertaining the suitability of the contents contained herein for a particular lab or commercial use, consultation with the subject expert is needed. In addition, while using the information and methods contained herein, the practitioners and researchers need to be mindful for their own safety, along with the safety of others, including the professional parties and premises for whom they have professional responsibility. To the fullest extent of law, the publisher, editor and authors are not liable in all circumstances (special, incidental, and consequential) for any injury and/or damage to persons and property, along with any potential loss of profit and other commercial damages due to the use of any methods, products, guidelines, procedures contained in the material herein.

A catalogue record for this book is available from the National Library of Australia

ISBN (print): 978-1-925823-70-7

Contents

Preface

Application of nanoparticles/nanomaterials for food and biomedical technology is a rapidly developing interdisciplinary research field. Rapid improvements in nanoscience and nanotechnology are prominent for the development of novel and feasible methods that may provide better quality of healthcare services. In this respect, the book "Recent Trends in Nanobiotechnology" provides insights into the recent advancements in nanobiotechnology and focuses on the applications of nanomaterials in food and biomedical sectors. The book begins with a chapter on the applications of the biodegradable nanocomposite films as materials for food packaging, followed by subsequent chapters on the recent advancements in nanobiosensors for detecting toxins and pathogens in food as well as nanomaterials based diagnostics. The contributors also illustrate the recent developments in nanomaterials for treatment of infectious diseases.

1

Applications of Biodegradable Nanocomposite Films for Food Packaging

Vikas S. Hakke,[a] Shirish H. Sonawane,[a,*] Irina Potoroko[b] and Irina Kalinina[b]

[a]Department of Chemical Engineering, National Institute of Technology Warangal, India
[b]Department of Food Technology and Biotechnology, South Ural State University, Russia

*Corresponding author: shirish@nitw.ac.in

1.1 Introduction

As an attractive branch of science, nanotechnology deals with crystals, crystal planes, atoms and molecules, or we can say with the small size in range of 100 Å to 10000 Å. Nanometric size of the material has novel properties as compared to its macro structure. High surface to volume ratio is the unique characteristic of the nanomaterials. Especially, nanocomposites have applications in areas like medicine, food technology, process industries and agriculture. Application of nanocomposite films in food processing and preservation is the fascinating topic for the researchers in the past few years.

Biopolymers react with enzymes or hormones generated by microorganisms to produce CO_2 and H_2O. The prefix bio indicates the biodegradability of such polymers. Biopolymers are increasingly becoming an alternative source to petroleum based synthetic plastics. Nanoparticles of different materials can be blended with biopolymers to enhance the properties. Different nanostructures such as nanocrystals, nanorods and nanofibers have distinct impact on the physical and mechanical properties of the polymers. In this respect, the biopolymers blended with nanoparticles lead to the development of biodegradable nanocomposites, which have extensive use in food processing and packaging, coatings, corrosion inhibition, etc. [1].

Recent Trends in Nanobiotechnology, edited by Prakash Saudagar and K. Divakar
© 2019 Central West Publishing, Australia

The food packaging films are required to enhance the shelf-life and safety of food products. The food quality is measured in terms of its physical appearance, color, odor and composition. The microbial activity within the food material is the main reason for the spoilage which cannot be stopped by any physical or chemical treatment without harming the purity of the food material. Thus, the shelf-life of the food is defined by the microbial activity associated with it. The microbial activity within the food material can be slowed down by avoiding the moisture (water) and oxygen contact of the food material, which can be effectively achieved by suitable packaging. Biodegradable nanocomposites have been used as barrier materials for food packaging applications. Multilayered films provide barrier properties at higher cost, thus, single layered nanocomposite films are needed to be develop so as to achieve high barrier performance at low cost.

Food packaging is mainly used to extend the quality of food during storage or transportation. Prevention of food from external damage, contamination by microbes and chemicals, moisture, light and oxygen are the required qualities of packaging materials/films. The microbial activity in the food varies with the moisture content, thus, it should obstruct the moisture transport. Also, it must provide resistance for permeation of gases, vapors, flavors and taints.

The major parameters of concern for the biodegradable nanocomposite films are their composition, physical appearance, strength, water and vapor permeability, time taken for film degradation, etc. In spite of the merits associated with biodegradable films, these also have several drawbacks due to their physical properties and difficulty in preparation. These drawbacks are the main hurdle to their use in food preservations application [2].

1.1.1 Synthetic and Natural Polymers used for Food Packaging

Currently, petrochemical based synthetic polymer packaging is being used and has its own pros and cons. High strength as well as effective resistance to water and gases are the main reasons for the use of synthetic polymers. High strength prevents the physical damage and gas barrier properties avoid the oxygen contact of the food, which arrests the microbial activity to some extent. This improves the shelf-life of the food. Though the synthetic polymers perform effectively, they are associated with disposal challenges due to non-degradability. Owing to covalent bands in the chains, the energy required for the degradation of petroleum-based polymers is much higher. Non-degradability of the packaging films leads to the accumulation of organic wastes in the environment. As a result, soil and air pollution as well as global warming phenomena are reaching alarming levels. Alteration in the composition of polymers films can offer a solution, like bio-based packaging.

Fragmentation or disintegration of natural polymers with the help of fungi, micro-organisms, oxygen, moisture and soil is the general process of biodegradation. Micro-organisms, through the enzymatic action, decompose biocomposites into biomass. In this respect, natural polymers are divided in three categories: i) biomass polymers or polymers extracted from biomass, for example polysaccharides, proteins, polypeptides, cellulose, etc., ii) chemically synthesized natural polymers like poly(lactic acid) and iii) micro-organisms synthesized polymers like pullan, bacterial cellulose, xanthan, etc.

Chemically synthesized biopolymers may present effective alternatives to synthetic polymers. Such nanostructured synthetic biopolymers offer good properties like high mechanical strength, barrier properties and biodegradability. Nanoscale particles can also be obtained from the natural materials like protein, cellulose, starch, lipids, etc., and can be blended with other nanoparticles like zein, collagen, nanoclays, etc., thus, leading to enhanced mechanical properties, permeability resistance to moisture and gases, durability, etc.

The introduction of nanocomposite films with embedded nanoparticles of different natural materials can be the strategy to deal with the above-mentioned issues of nanocomposite films. The cost of production of the nanocomposite films is also a major issue as it restricts widespread application [3-5].

1.1.2 Nanotechnology in Food Processing

Food matrix shows better performance in taste, texture and consistency when they are associated with the nanostructured food ingredients [6]. Incorporation of nanotechnology in food preservation through the prevention of micro-organism growth on the food matrix helps to avoid the spoilage of food. Thus, the shelf life of food gets increased during storage and transportation [7]. Due to very small structural size, nanoparticles can be utilized as additive carriers in the food matrix without hampering the morphology of the food matrix. These nano-carriers can be utilized for the delivery of additives such as aroma, flavor, etc., which helps to improve the food quality. Sometimes, bioactive compounds can also be delivered at various sites within the human body. As the body cells only absorb submicron nanoparticles size, thus, this application of nano-carriers is much useful [8]. The nano-carriers for the above function must possess some properties: i) these should have affinity towards the carried material, i.e. the ability to maintain the carried material for long time period, and ii) the release characteristics of the carried materials and the ability to deliver at the target should be defined. Nano-encapsulated materials, emulsions, biopolymers and colloids are the commonly used application forms of nano-carriers. Nanocomposites developed using such methods are suitable options for the replacement of synthetic films as such nanocomposites can alter the basic properties of biopolymers. Some of the nanocomposites have also suitability for detection of pathogens and microbes, thus, preventing food from contamination [9].

Nanostructured food ingredients in food matrix not only improve the shelf life of the food, but also help to maintain the nutritional value of the food, by helping to maintain the aroma, flavor and appearance. Improvement of the nutritional value of the food matrix can also be achieved with the help of nano-carriers, which helps to develop nutritional food matrix with improved shelf life.

Packaging films possessing good mechanical strength, oxygen barrier characteristics and anti-microbial activity are the ground-breaking concepts in food packaging. Non-toxicity of nanocomposites must be considered when used in film packaging applications. Encapsulation of the nanoparticles like natural anti-microbial agents (e.g. Nisin), metals (e.g. silver) and metal oxides (e.g.TiO_2) are useful to improve the oxygen and microbial barrier as well as mechanical properties of the biodegradable films. Cellulose acetate (CA), polyethylene glycol (PEG), lactic acid, starch and proteins are used to encapsulate the nanoparticles for nanocomposite films for food packaging.

1.1.3 Micro- and Nano-encapsulation Techniques

Encapsulation is the method of capturing solid particles, liquid droplets and gas bubbles in solid or liquid envelop made up of other immiscible material. Two types of encapsulation systems are core/shell and particle/matrix, as shown in Figure 1.1.

In the first case, continuous phase core material is surrounded by the membrane shell. On the other hand, in the second case, micro- or nanoparticles (solid or liquid droplets) form a dispersed phase surrounded by the continuous liquid or solid phase. In both cases, the encapsulated materials (core) are protected by the surrounding layer (shell), so that the protected inner material can be released as per the need.

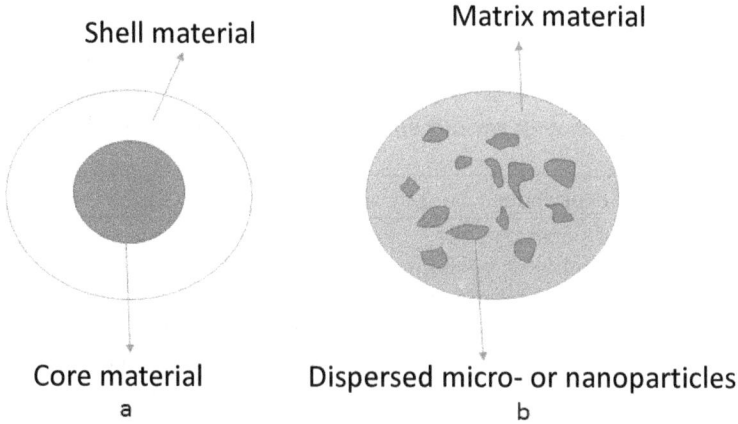

Shell material   Matrix material

Core material   Dispersed micro- or nanoparticles

a   b

Figure 1.1 Encapsulation methods: (a) core/shell and (b) particles in matrix.

The foremost profits of nano-encapsulation for the food matrix is the improvement in aroma, flavor ingredients, nutritional components like vitamins, carbohydras, protein, minerals, etc. Incorporation of encapsulation techniques provides controlled interaction of the ingredients within the food matrix, improved soluble gas interaction, prevention of contamination from pathogen or microorganisms, improved solubility of the poorly water soluble ingredients [10,11].

Controlled release of active ingredient at the target food matrix, protection against the pathogens and taste enrapturing are the most fascinating application of the encapsulation techniques. Traditional nanoparticles have difficulties of interaction with food matrix, while encapsulation provides an improved environment, and the controlled release makes them more effective in delivery systems. Encapsulation of the nanoparticles keeps them away from the environment of the food matrix, which allows efficient release of the active component at the intended area of food matrix as well as body [12]. Table 1.1 presents the comparison of various encapsulation techniques [13].

The successful attempt to deliver the encapsulated aroma molecules in halloysites nano-containers shows that there is a possibility of the development of food matrix where volatile aroma molecules can be stored for longer period of time. The development of nano-containers responsive to enzymes, pH of surrounding, heat and pressure can be utilized in target delivery. For example, hollow lumen can be used to capture volatile aroma molecules. The nano-container hollow lumen coated with polyelectrolytic coating is reported to be sensitive to the pH level of the surroundings in range of 3-7 [14].

Table 1.1 Comparison of various encapsulation techniques [13]

Technique	Properties	Examples
Edible coatings	To preserve the quality of fresh food during extended storage	Gelatin-based edible coatings containing cellulose Nano-crystals Chitosan/nano-silica coatings Chitosan films with nano-SiO_2 Alginate/lysozyme nano-laminate coatings
Hydrogels	Can be easily placed into capsules; protects drugs from extreme environments	Protein hydrogels, poly(acrylic acid) (PAA) and polyacrylamide (PAM) nanocomposite hydrogels,
Nano-emulsions	Greater stability to droplet aggregation and gravitational separation; higher optical clarity; increased oral bioavailability	Anti-oxidant essential oil encapsulated with protein makes stable nano-emulsion with water

1.2 Encapsulation Technologies for Preserving Properties of Bioactive Substances in Food Matrix

Nowadays, there is an urgent need to introduce new approaches directed at the formation of functional food matrix. Under conditions of food shortage and health problems, the preservation of macro- and micro-nutrients in food is of great importance. It is vitally important to obtain the products capable not only of satisfying the needs of a modern age towards food processing substances and energy, but also to provide the body with necessary micro-nutrients without modification. Clearly, the creation of new systems of delivering the functional food ingredients is needed, which will make it possible to stabilize the substance in the product system and to ensure the release at the target. Here, micro- and nano-encapsulation technologies are of particular interest. Such approach for obtaining enriched food products allows us to exclude unpredictable changes in the food matrix of the product and provide delivery vector of a bio-active component.

At present, there exist some approaches for effectively designing the food matrix. However, addition of new substances to the product does not always lead to synergy and may result in substance incompatibility with product nutrients. To overcome this problem, one can apply the system of capsule formation, where a new substance (nutraceutical) can be placed inside the capsule, important owing to its therapeutic effect (anti-oxidants, vitamins, amino acids, etc.). In order to improve the nutraceutical bio-availability, one can apply micro-emulsions, nano-emulsions, multi-phase emulsions, nano-dispersed solutions, microgels, etc. [9,12,16]. Each of these systems has its own advantages and disadvantages, thus, the choice of a delivery system suitable for a certain application is a multi-factorial task. To form the sustainable system of micro- and nano-capsules loaded with nutrients, one should take into account the compatibility of the shell and the substance inside it. The examples of successful methods are presented in Table 1.2.

Table 1.2 Methods used for nutraceutical encapsulation

Nutraceutical	Encapsulation method	Result	Source
Resveratrol	Adding glycosyl group to resveratrol via a succinate linker	Water solubility increase	Biasutto *et al.* (2009)
Taxifolin	Obtaining conjugates based on cyclodextrins	Solubility and thermal stability increase	Yang *et al.* (2011)
Anthocyanins	Encapsulation in polysaccharide microcapsules	Thermal stability increase and preservation of anti-oxidant properties	Chen Tan *et al.* (2018)
Quercetin	Conjugates of amino acids of synthesized quercetin (e.g. quercetin-glutamic acid conjugate)	Water solubility increases and increase of cell permeability in comparison to the original molecule	Kim *et al.* (2009)

Curcumin	Obtaining conjugates based on cyclodextrins	Anti-inflammatory bio-availability	Rocks *et al.* (2012)
Taxifolin	Obtaining nano-emulsion based on corn oil	Bio-availability and bio-activity increase	Potoroko *et al.* (2018)

The most effective way of preserving encapsulated substance is achieved by modeling based on quantum chemical calculations. Multi-conformational and multi-tautomeric analysis of compounds helps to predict the encapsulation mechanism, transportation and release of bio-active substance, its behavior in the body, biological activity in cell targets, etc. Thus, modeling taxifolin complexes with beta-cyclodextrin (Figure 1.2) vividly describes the mechanism of bio-active substance formation and enables to predict its properties, solubility in particular. Taxifolin is well incorporated into beta-cyclodextrin tube by means of dihydroxyphenyl substituent which adjoins densely to the atoms of the inner part of the tube, thus, forming a series of hydrophobic interactions with the CH-fragments of the monomer units of glucose. Besides, hydrophilic OH-groups of the dihydroxyphenyl substituent are located outside the tube which makes hydration possible, thus, increasing the DKG solubility in polar solvents.

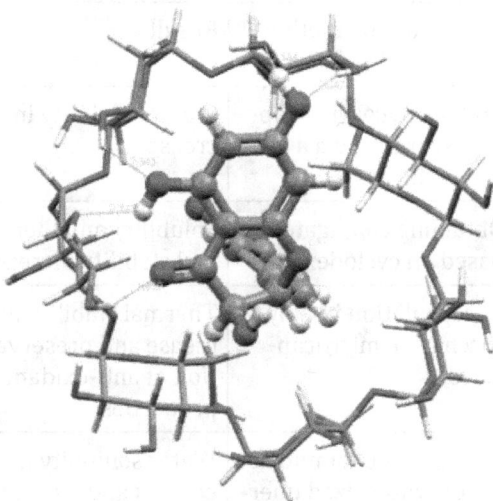

Figure 1.2 Model of taxifolin and beta-cyclodextrin complex formation (using MOPS algorithm) [15,22].

On the whole, the encapsulation systems must meet general requirements which include: compatibility with food product, reliability, stability towards environmental exposure, economic viability, etc. Studies in the field of micro- and nano-encapsulation of nutaceuticals often show favorable results for achieving goals as increasing nutraceutical solubility, bio-availability and bio-activity.

1.3 Characteristics of Nanocomposite Films

The synthetic petroleum-based plastics have relatively high mechanical, thermal and barrier properties as compared to the biodegradable polymer films, which make them favorable for packaging applications. Nanocomposites of natural materials are suitable to overcome the deficiencies in the biodegradable natural polymers, thereby, competing with the synthetic polymers. The biodegradable natural polymers have properties like minimal branching, low crystallinity, low degree of polymerization and water solubility which must be improved for use in food packaging applications. Most important properties of the packaging film based on biopolymers are non-toxic nature, resistance to pathogens and moisture, etc.

Mechanical strength of bio-nanocomposite materials is basically achieved due to the nanoscale particles which improve the elastic modulus of the polymer. Overall, the reinforcement phase at nanoscale dimensions provides improvements in terms of chemical resistance, gas and moisture permeability, electrical and optical properties, etc. [23,24]. Addition of nanofillers in biopolymers also helps to modulate effective films for packaging applications which keep the food matrix isolated from the deteriorative surroundings. The diffusion of the gases through the pores of the biopolymer films depends on the crystallinity of the polymer system. Crystalline system represents compactness of the polymer which comes from the cohesive forces between molecules or crystals, which provide them additional cohesive strength. While polar functional groups moderate molecular self-diffusion, strongly active polar groups or hydrogen bonding will generate the ordered chains. The intermolecular forces, which promote cohesion (particularly hydrogen bonding), enhance the crystallinity. As the intermolecular distance is less, polar groups are near to each other, thus, promoting the chemical and physical bonding which helps to enhance the crystallinity.

The films casted from natural polymers exhibit different characteristics as the casting method is changed. The structural randomness varies with respect to the method of casting. This randomness in the monomer arrangement affects the strength of the films. Cast film properties like stiffness, yield point, permeability, flexibility and brittleness also depend on the crystallinity of the films [25].

Additional release studies of encapsulated materials will help to develop active and smart packaging films, which can be utilized for the detection of damaged cell in body, smart target delivery of drugs in the body or nutrition in food matrix [26].

Anti-bacterial property is one of the most essential properties which affect the shelf life of the food matrix. The blending of nanofillers as an active material for packaging films can be used to improve anti-bacterial activity [27]. For instance, the anti-bacterial properties of the packaging films are observed to improve with cupper nanofillers in cellulose films, along with providing strength to cellulose films. Nanoscale particles of organic compounds like essential oil, organic acids and bacterial enzymes are also used as active material to blend in biopolymers [28,29].

Modifications of the natural polymers are needed to make them suitable for food packaging. The nanocomposites of more than two natural polymers are also needed to be studied. The polymers like poly(lactic acid), polysaccharides, proteins and polyesters have differential compatibility with each other and have the ability to form crystalline structured films. Melt intercalation method using extrusion or injection molding has considerably low-cost investment which can be commercialized for the bulk production of nanocomposite of natural polymers. The thermal degradation of polymer is the main process requiring optimization in melt intercalation.

Many of the active materials are sensitive to temperature and pressure, are sometimes not suitable for food composition. Use of inorganic material nanoparticles as nanofillers like Ag, Cu and Au results in effective anti-bacterial active ingredient [30]. Therefore, in recent years, the use of inorganic materials in food packaging films has seen an increasing trend to improve the anti-microbial activity of films [31,32]. Microbial growth is completely hindered due to the effective addition of the anti-microbial organic and inorganic nanofillers, which prevents food spoilage, thus, leading to enhanced shelf-life.

The properties of the food packaging films are not only restricted to the mechanical performance, anti-microbial activity, smart release of the encapsulated material in the film matrix and barrier properties, but these also extend with special requirements of the food matrix.

For instance, the requirements of shock-proof films, improved transparency, flavor and aroma masking, etc., still need to be worked out. Overall, most of these requirements can be fulfilled by different organic or inorganic nanofillers.

Zinc oxide (ZnO) nanoparticles used as filler in organic gelatin prepared from fish have been reported to exhibit significant improvement in the optical properties, UV protection activity and mechanical strength of the films at low filler concentrations. However, the effect of ZnO on the microbial activity of the gelatin-based films was not remarkable. Therefore, ZnO based nanocomposite biofilms have excellent applications in food packaging and UV shielding [33].

Figure 1.3 also presents food packaging approaches and their categories.

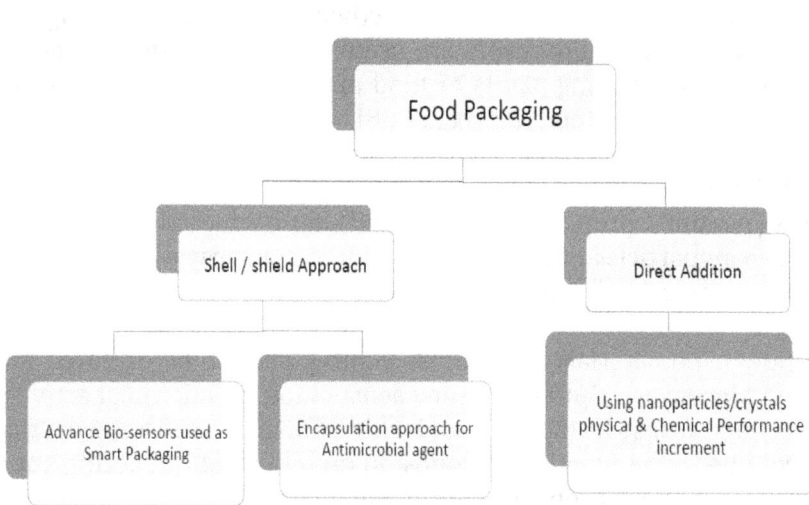

Figure 1.3 Food packaging approaches and their categories.

1.3.1 Anti-oxidant Nanocomposite Films

It is observed that the polymer crystals formed with the polar groups are more ordered and non-porous due to effective intermolecular interactions. On the other hand, less polar films are more porous, thus, permitting the permeation of oxygen, but not necessarily the larger water molecules. Microbial growth on the food matrix depends on the oxygen contact and moisture availability within the matrix. Thus, the spoilage of food depends on the oxygen permeability of the film used for packaging. Lipid oxides are responsible for the rancid odors and flavor of the food material, along with decreased nutritional value of the food [34]. Films generated from protein and carbohydrates show excellent resistance to oxygen because of their tightly packed crystalline structure [35].

1.3.2 Nanosensors for Pathogen Detection

The target delivery studies show that the pathogens can be detected by small changes in the enzymatic concentration on the food or film surface. The detailed studies show that nano-encapsulated fillers can be used as biosensors, where their release in the food matrix self-indicates the presence of pathogens [36]. Such biosensors offer very high level of sensitivity as they are activated with a small change in the composition. Nanosensors are useful for the detection of pathogens in the processing plants or food materials, thus, providing the safety aspect to the food matrix [37,38].

Development of functional nanosensors with the inorganic nanoparticles (electronic, optical and magnetic) blended with biopolymers (peptides, proteins and lactic acid) has been carried out. Some of the nanoparticles-based sensors include acoustic wave biosensors, optical biosensors, magnetic and electrochemical biosensors, etc. [39].

New nanomaterials and nanostructures need to be explored for use in biosensors. Optical immune sensors for the microbial activity detection can also be incorporated in thin films. This sensor emits the signal due to the functional groups on target cell [40]. Specific antibodies immobilized on the porous nanoparticles of alumina membranes help to detect the waterborne pathogens [41]. The specific chemical environment of pesticides [42], pathogens [43] and toxicity [44] can also be detected by the nanocomposites-based tracing biosensors.

1.3.3 Anti-microbial Food Packaging

Different types of nanofillers are used to the produce bio-nanocomposite films with enhanced properties for packaging applications. ZnO is observed to be useful for improving the thermal stability, UV protection and anti-microbial activity [45]. Silver nanoparticles (AgNPs), used since ancient times in the food matrix, improve the nutritional value as well anti-bacterial strength of the food matrix.

AgNPs when blended with the biopolymer films generate more compact structure, thus, leading to improved water vapor barrier and surface hydrophobicity [46].

1.4 Preparation Methods of Nanocomposite Films

Preparation of nanoparticles is achieved through two different approach, one is top down approach in which the conversion of macro-level to micro-level, followed by intensive size reduction to nano-level is carried out. Another way to generate the nanoparticles from their basic constituents by restricting the crystal formation to nanometer size using reduction or oxidation reactions is generally known as bottom up process. Both top down and bottom up processes are used depending up on the ease of generating the nanoparticles, the only difference is the nanoparticles generated with bottom up processes have uniform crystal structure.

Blending of two or more nanoparticles to obtain a uniform mixture is the prime objective of various methods. Nanoparticles used as nanofillers are generally chosen on the basis of the cost of process, material cost and ease of synthesis.

Polymer assisted fabrication of nanomaterials has been reported to be an effective technique for nanofiller synthesis. The intermolecular interaction between the nanofillers plays an important role in the development of the nanocomposite films. Properties like resistance to moisture, percolation, gas diffusion and strength are dependent on the intermolecular interactions. Such intermolecular interactions generate the cohesive forces which define the film nature. The interactions between nanofillers are explained by theories of van der Waals forces, hydrogen bonding, ionic bonding, colloidal and hydrophobic interactions, etc. Different technologies are used to generate the stable nanocomposite films [47]. Release of nanofillers in biopolymers can be achieved by different methods through blending, dispersion, surface modification, solution deposition, etc. [48]. Cost-effective, ecofriendly and multifunctional ergonomics of the product are required in the final bio-nanocomposite films produced by any of these methods. Few methods of nanocomposites synthesis are compared in Table 1.3.

1.5 Limitations of Nanocomposite Films and Future Scope

Development of bio-nanocomposites and their use in food packaging

Table 1.3 Synthesis methods for nanocomposite films

Method	Process description
Blending method	Mixing is frequently adopted for inclusion of ex-situ nanostructured materials into biopolymers. The nanoparticles with different crystal structures can be blended with the biopolymers. Nanofibers, nanorods, nanotubes and nanosheets are generally blended with the biopolymers. The interphase or intermolecular interaction define the stability of the blended nanocomposites. High shear rate of mixing is recommended for de-agglomeration of nanoparticles. Blending can be achieved through sonication, high pressure homogenization, microfluidic techniques, milling, etc.
Surface modification of nanocomposite	Dispersion of nanoparticles in this method is achieved due to the compatibility with biopolymer matrix. Chemical attachment of the nanoparticles with the polymer chains increases the strength of the polymer as well reduces the chances of agglomeration. Chemical and physical bonding of polymer and nanostructured particles require special techniques, like grafting. In most of cases, the functional groups play an important role in attachment. Nanoparticles attachment is classified as "grafting to" and "grafting from". The hydroxyl functional groups of the nanofillers are generally attached to the polymers in "grafting to" approach, while in "grafting from" approach, the polymer extension or propagation is achieved from the functional groups on the filler surface.
Dispersion	Dispersion of nanoparticles in solvent is carried out in this method, e.g. oil in water emulsion, essential oil encapsulation in protein, etc. The method is useful in many applications due to low cost, and the films with stable emulsions exhibit good strength. The encapsulated materials help to alter the properties of the films with smart release. Encapsulation of many food ingredients can be achieved through the dispersion technique. For example, the encapsulation of aroma in halloysite coated with electrolyte is helpful for the storage and sensitive release of volatile aroma molecules [14]. Spray drying is another example of dispersion in which the continuous matrix substrate carries the core material. Water is most commonly used as the continuous phase.

applications has a positive impact on the environment, as there is no accumulation of non-degradable polymer waste. Application of nanocomposites have several limitations when compared with conventional synthetic polymers. The issues like environmental impact, consumer safety, policies and regulations regarding the use of biopolymers are required to be resolved. The profit-making approach of adding low cost inorganic materials may result in health issues. The nanocomposite biofilms have potential for extensive application, but their effect on human body is yet to be studied. The nanoscale material blending has both positive as well as negative impacts. As a typical example involving the blending of Ag nanoparticles, aggregation is the major problem due to the intermolecular interactions of the nanoparticles of different size and shape [49]. Silver nanoparticles have high surface area and ability to penetrate the cell wall, due to which they act as good anti-bacterial agents. However, the same silver nanoparticles in the large scale (macro) exhibit toxicity towards the body cells [50,51]. The agglomeration exhibits unstable form of the nanocomposites, which must be handled carefully. Contamination by agglomerated nanoparticles in the food packaging films considerably affects their use on biological cells [52].

The impact of the nanosized particles on the respiratory system, porous skin and eyes are yet to be studied. As the nanoparticles exhibit different behaviors at nano-level as compared to their macrostructure, thus, such studies become very important [53]. On degradation of the nanocomposite films in the environment, these fillers may affect the environmental cycle which can causes various issues. Due to the extremely small size, nanoparticles have the ability to penetrate or diffuse through the porous materials. The tendency bond with the food matrix can be disadvantageous in the environmental cycle. In terms of safety of the environmental cycle, it is necessary to evaluate the potential hazards and their assessment. Impact of the ingredient of the nanocomposites on the surrounding plants and organisms must also be analyzed [54]. Also, the mixing of the nanoparticles from the biopolymer packaging film to the packed food is another important concern.

Non-degradable materials have been used in many instances for the food encapsulation which do not meet the environmental safety requirement [55]. Controlled release studies of the encapsulated ingredients in the food matrix, along with their compatibility during storage and functionality in the packaging films play an important role [56]. The encapsulated ingredients must be active till their full utilization in the packaging films. For instance, the encapsulation of essential oil in protein nanoparticles can be incorporated in starch to enhance the anti-oxidant properties of the films. New policies are also needed to be formed for the safety of consumers. Every consumer must be alert about the merits and demerits of the encapsulated food packaging [57]. Need of regulatory guidance to consumers through proper channels must also be organized for the safe commercialization of nanotechnology for food packaging applications [58]. Furthermore, it is advised that the nanoparticle toxicity data should be open to the public in terms of both bulk and surface properties [59].

1.6 Conclusions and Future Aspects

The prominent sectors like pharmaceutical industries, medical and food technology are impacted by the promising applications of nanocomposites. Smaller size of the nanoparticles leads to superior properties than the macro-structure, due to which the bio-nanocomposites find many diverse applications, which also lower the carbon footprint. In this chapter, we have discussed the applications of nanoparticles in food packaging. Natural polymers offer suitable alternative to the synthetic polymers due to ecofriendly and biodegradable nature. The properties of the biodegradable natural polymers can be improved by incorporating nanoparticles. As a result, the bio-nanocomposite films exhibit improvements in mechanical strength, oxidation resistance, gas and moisture barrier, etc. The use of active nanoparticles, depending up on their functional groups, as bio-sensors also presents a strong potential for food preservation as well as to detect the pathogenic activity in food and surroundings. Nanofillers can also be used in food packaging as anti-microbial agents, gas sensors, temperature sensors and oxygen scavengers. The addition of nanoparticles such as essential oil, silver and TiO_2 to the biopolymers such as lactic acid and starch opens up numerous opportunities to develop novel nanocomposite materials with smart and sensitive applications, such as food packaging.

References

1. Hernandez, R. J., Selke, S. E. M., and Culter, J. D. (2000) Major plastics in packaging. In: *Plastics Packaging: Properties, Processing, Applications and Regulations*, Carl Hanser Verlag, Germany.

2. Brooks, D. (2000) Types of plastics materials, barrier properties and applications. In: *Materials and Development of Plastics Packaging for the Consumer Market*, Giles, G. A., and Bain, D. R. (eds.), Sheffield Academic Press, UK, pp. 16-45.

3. Qureshi, A. M., Karthikeyan, S., Punita, K., Khan, P. A., Urpit, S., and Mishra, U. K. (2012) Application of nanotechnology in food and dairy processing: An overview. *Pakistan Journal of Food Science*, **22**, 23-31.

4. Pradhan, N., Singh, S., Ojha, N., Srivastava, A., Barla, A., Rai, V., and Bose, S. (2015) Facets of nanotechnology as seen in food processing, packaging, and preservation industry. *BioMed Research International*, **2015**, doi: 10.1155/2015/365672.

5. Gupta, A., Eral, H. B., Hatton, T. A., and Doyle, P. S. (2016) Nanoemulsions: formation, properties and applications. *Soft Matter*, **12**, 2826-2841.

6. *Nanotechnologies in the Food Industry* (2006). Online: http://www.cientifica.com/www/details.php?id47 [accessed 24th October 2006].

7. Radha, K. Thomas, A., and Sathian, C. T. (2014) Application of nano technology in dairy industry: prospects and challenges - A review. *Indian Journal of Dairy Science*, **67**(5), 367-374.

8. Ezhilarasi, P. N., Karthik, P., Chhanwal, N., and Anandharamakrishnan, C. (2013) Nanoencapsulation techniques for food bioactive components: A review. *Food Bioprocess Technology*, **6**, 628-647.

9. Bratovcic, A., Odobašic, A., Catic, S., and Šestan, I. (2015) Application of polymer nanocomposite materials in food packaging. *Croatian Journal of Food Science and Technology*, **7**, 86-94.

10. Ubbink, J., and Kruger, J. (2006) Physical approaches for the delivery of active ingredients in foods. *Trends in Food Science and Technology*, **17**, 244-254.

11. Weiss, J., Takhistov, P., and McClements, J. (2006) Functional materials in food nanotechnology. *Journal of Food Science*, **71**, R107-R116.

12. Lamprecht, A., Saumet, J. L., Roux, J., and Benoit, J. P. (2004) Lipid nanocarriers as drug delivery system for ibuprofen in pain treatment. *International Journal of Pharmaceutics*, **278**, 407-414.

13. Singh, T., Shukla, S., Kumar, P., Wahla, V., Bajpai, V. K., and Rather, I. A. (2017) Application of nanotechnology in food science: Perception and overview. *Frontiers in Microbiology*, doi: 10.3389/fmicb.2017.01501.

14. Ghodke, S. A., Sonawane, S. H., Bhanvase, B. A., Mishra, S., and Joshi, K. S. (2015) Studies on fragrance delivery from inorganic nanocontainers: encapsulation, release and modeling studies. *Journal of the Institution of Engineers (India): Series E*, **96**(1), 45-53.

15. Grishina M., Bolshakov O., Potemkin A., and Potemkin V. (2016)

Theoretical investigation of electron structure and surface morphology of titanium dioxide anatasenano-particles. *Computational and Theoretical Chemistry*, **1091**, 122-136.

16. Tan, C., Selig, M. J., Lee, M. C., and Abbaspourrad, A. (2018) Encapsulation of copigmented anthocyanins within polysaccharide microcapsules built upon removable CaCO3 templates. *Food Hydrocolloids*, **84**, 200-209.

17. Yang, L.-J., Chen, W., Ma, S.-X., Gao, Y.-T., Huang, R., Yan, S.-J., and Lin, J. (2011) Host-guest system of taxifolin and native cyclodextrin or its derivative: Preparation, characterization, inclusion mode, and solubilisation. *Carbohydrate Polymers*, **85**(3), 629-637.

18. Biasutto, L., Marotta, E., Bradaschia, A., Fallica, M., Mattarei, A., Garbisa, S., Zoratti, M., and Paradisi, C. (2009) Soluble polyphenols: Synthesis and bioavailability of 3,4′,5-tri(α-d-glucose-3-O-succinyl) resveratrol. *Bioorganic and Medicinal Chemistry Letters*, **19**(23), 6721-6724.

19. Kim, M. K., Park, K,-s., Yeo, W.-s., Choo, H., and Chong. Y. (2009) In vitro solubility, stability and permeability of novel quercetin-amino acid conjugates. *Bioorganic and Medicinal Chemistry*, **17**(3), 1164-1171.

20. Rocks, N., Bekaert, S., Coia, I., Paulissen, G., Gueders, M., Evrard, B., Van Heugen, J.-C., Chiap, P., Foidart, J.-M., Noel, A., and Cataldo, D. (2012) Curcumin-cyclodextrin complexes potentiate gemcitabine effects in an orthotopic mouse model of lung cancer. *British Journal of Cancer*, **107**(7), 1083-1092.

21. Potoroko, I., Kalinina, I. V., Naumenko, N. V., Fatkullin, R. I., Shaik, S., Sonawane, S. H., Ivanova, D., Kiselova-Kaneva, Y., Tolstykh, O., and Paymulina A. V. (2018) Possibilities of regulating antioxidant activity of medicinal plant extracts. *Human Sport Medicine*, **17**(4), 77-90.

22. Potemkin, V., and Grishina, M. (2018) Electron-based descriptors in the study of physicochemical properties of compounds. *Computational and Theoretical Chemistry*, **1123**, 1-10.

23. Pathakoti, K., Manubolu, M., amd Hwang, H.-M. (2017) Nanostructures: Current uses and future applications in food science. *Journal of Food and Drug Analysis*, **25**, 245-253.

24. Rhim, J. W., and Ng, P. K. (2007) Natural biopolymer-based nanocomposite films for packaging applications. *Critical Reviews In Food Science and Nutrition*, **47**, 411-433.

25. Couch, L. M., Wien, M., Brown, J. L., and Davidson, P. (2016) Food nanotechnology: proposed uses, safety concerns and regulations. *AgroFood Industry Hi Tech*, **27**, 36-39.

26. Mihindukulasuriya, S. D. F., and Lim, L. T. (2014) Nanotechnology development in food packaging: A review. *Trends in Food Science and Technology*, **40**, 149-167.

27. Pinto, R. J. B., Daina, S., Sadocco, P., Neto, C. P., and Trindade, T.

(2013) Antibacterial activity of nanocomposites of copper and cellulose. *BioMed Research International,* **2013,** doi: 10.1155/2013/280512

28. Gálvez, A., Abriouel, H., López, R. L., and Omar, N. B. (2007) Bacteriocin-based strategies for food biopreservation. *Journal of Food Microbiology,* **120,** 51-70.
29. Schirmer, B. C., Heiberg, R., Eie, T., Møretrø, T., Maugesten, T., and Carlehøg, M. (2009) A novel packaging method with a dissolving CO_2 headspace combined with organic acids prolongs the shelf life of fresh salmon. *Journal of Food Microbiology,* **133,** 154-160.
30. Soares, N. F. F., Silva, C. A. S., Santiago-Silva, P., Espitia, P. J. P., Gonçalves, M. P. J. C., Lopez, M. J. G., Miltz, J., Cerqueira, Miguel A., Vicente, A. A., Teixeira, J. A., Silva, W., Botrel, D. (2009) Active and intelligent packaging for milk and milk products. In: *Engineering Aspects of Milk and Dairy Products,* Coimbra, J. S. R., and Teixeira, J. A. (eds.)., CRC Press, USA, pp. 155-174.
31. Bradley, E. L., Castle, L., and Chaudhry, Q. (2011) Applications of nanomaterials in food packaging with a consideration of opportunities for developing countries. *Trends Food Science and Technology,* **22,** 603-610.
32. Tan, H., Ma, R., Lin, C., Liu, Z., and Tang, T. (2013) Quaternized chitosan as an antimicrobial agent: antimicrobial activity, mechanism of action and biomedical applications in orthopedics. *Internaitonal Journal of Molecular Science,* **14,** 1854-1869.
33. Yanishieva, N. N. V., Marinova, E., and Pokorny, J. (2006) Natural antioxidant from herbs and spices. *European Journal of Lipid Science and Technology,* **108,** 776-793.
34. Yang, L., and Paulson, A. T. (2000) Effect of Lipids on mechanical and Moisture barrier properties of edible gellan film. *Food Res. Int.,* **33,** 571-578.
35. Cheng, Q., Li, C., Pavlinek, V., Saha, P., and Wang, H. (2006) Surface-modified antibacterial TiO_2/Ag^+ nanoparticles: preparation and properties. *Applied Surface Science,* **252,** 4154-4160.
36. Helmke, B. P., and Minerick, A. R. (2006) Designing a nano-interface in a microfluidic chip to probe living cells: challenges and perspectives. *Proceedings of National Academy of Sciences of the USA,* **103,** 6419-6424.
37. Bouwmeester, H., Dekkers, S., Noordam, M. Y., Hagens, W. I., Bulder, A. S., Heer, C., ten Voorde, S.E.C.G., Wijnhoven, S. W. P., Marvin, H. J. P., and Sips, A. J. A. M. (2009) Review of health safety aspects of nanotechnologies in food production. *Regulatory Toxicology and Pharmacology,* **53,** 52-62.
38. Jianrong, C., Yuqing, M., Nongyue, H., Xiaohua, W., and Sijiao, L. (2004) Nanotechnology and biosensors. *Biotechnology Advances,* **22,** 505-518.

39. Subramanian, A. (2006) A mixed self-assembled monolayer-based surface Plasmon immunosensor for detection of E. coli O157:H7. *Biosensors and Bioelectronics*, **7**, 998-1006.

40. Tan, F., Leung, P. H. M., Liud, Z., Zhang, Y., Xiao, L., Ye, W., Zhang, X., Yi, L., and Yang, M. (2011) microfluidic impedance immunosensor for E. coli O157:H7 and Staphylococcus aureus detection via antibody-immobilized nanoporous membrane. *Sensors and Actuators B: Chemical,* **159**, 328-335.

41. Liu, S., Yuan, L., Yue, X., Zheng, Z., and Tang, Z. (2008) Recent advances in nanosensors for organophosphate pesticide detection. *Advanced Powder Technology*, **19**, 419-441.

42. Inbaraj, B. S., and Chen, B. H. (2015) Nanomaterial-based sensors for detection of foodborne bacterial pathogens and toxins as well as pork adulteration in meat products. *Journal of Food and Drug Analysis,* **24**, 15-28.

43. Palchetti, I., and Mascini, M. (2008) Electroanalytical biosensors and their potential for food pathogen and toxin detection. *Analytical and Bioanalytical Chemistry,* **391**, 455-471.

44. Kanmani, P., and Rhim, J.-W. (2014) Properties and characterization of bionanocomposite films prepared with various biopolymers and ZnO nanoparticles. *Carbohydrate Polymers*, **106**, 190-199.

45. Rhim, J. W., Wang, L. F., and Hong, S. I. (2013) Preparation and characterization of agar/silver nanoparticles composite films with antimicrobial activity. *Food Hydrocolloids,* **33**, 327-335.

46. Rozenberg, B. A., Tenne, R. (2008). Polymer-assisted fabrication of nanoparticles and nanocomposites. *Progress In Polymer Science*, **33**(1), 40-112.

47. Rouhi, J., Mahmud, S., Naderi, N., Raymond Ooi, C. H., and Mahmood, M. R. (2013) Physical properties of fish gelatin-based bio-nanocomposite films incorporated with ZnO nanorods. *Nanoscale Research Letters*, **8**, 1-6.

48. Poortavasoly, H., Montazer, M., and Harifi, T. (2014) Simultaneous synthesis of nano silver and activation of polyester producing higher tensile strength aminohydroxylated fiber with antibacterial and hydrophilic properties. *RSC Advances*, **4**, 46250-46256.

49. Alexandre, M., and Dubois, P. (2000) Polymer-layered silicate nanocomposites: preparation, properties and uses of a new class of materials. *Materials Science and Engineering R: Reports*, **28**(1-2), 1-63.

50. Kornmann, X., Lindberg, H., Berglund, L. A., 2001b. Synthesis of epoxy-clay nanocomposites: influence of the nature of the clay on structure. *Polymer*, **42**(4), 1303-1310.

51. Liu, L., Liu, J. C., Wang, Y. J., Yan, X. L., and Sun, D. D. (2011) Facile synthesis of nanodispersed silver nanoparticles on grapheme oxide sheets with enhanced antibacterial activity. *New Journal of Chemistry*, **35**, 1418-1423.

52. Cushen, M., Kerry, J., Morris, M., Cruz-Remero, M., and Cummins, E. (2014) Evaluation and simulation of silver and copper nanoparticle migration from polyethylene nanocomposites to food and an associated exposure assessment. *Journal of Agricultural and Food Chemistry*, **62**, 1403-1411.

53. Marambio-Jones, C., and Hoek, E. M. V. (2010) A review of the antibacterial effects of silver nanomaterials and potential implications for human health and the environment. *Journal of Nanoparticle Research*, **12**, 1531-1551.

54. Carlson, C., Hussain, S., Schrand, A., Braydich-Stolle, L., Hess, K., Jones, R., and Schlager, J. (2008) Unique cellular interaction of silver nanoparticles: sizedependent generation of reactive oxygen species. *The Journal of Physical Chemistry B*, **112**, 13608-13619.

55. Sozer, N., and Kokini, J. (2012) *The Applications of Nanotechnology*, Elsvier, USA.

56. Othman, S. H. (2014) Bio-nanocomposite materials for food packaging applications: type of biopolymer and nano-sized filler. *Agriculture and Agricultural Science Procedia*, **2**, 296-303.

57. Augustin, M. A., and Heman, Y. (2009) Nano- and micro-structured assemblies for encapsulation of food ingredients. *Chemical Society Reviews*, **38**, 902-912.

58. Mills, A., and Hazafy, D. (2009) Nanocrystalline SnO2-based, UVB-activated, colourimetric oxygen indicator. *Sensors and Actuators B: Chemical*, **136**(2), 344-349.

59. Elzey, S., Larsen, R. G., Howe, C., and Grassian, V. H. (2009) Nanoscience and nanotechnology: Environmental and health impacts. In: *Nanoscale Materials in Chemistry*, Klabunde, K. J., and Richards, R. M. (eds.), 2nd edition, John Wiley and Sons, USA, pp. 681-727.

2

Nanotechnology Developments in Food Packaging

K. Radha Krishnan,[1]* Prakash Kumar Nayak,[1] S. Babuskin[2] and C. Chandramohan[3]

[1]*Department of Food Engineering and Technology, CIT, Assam, India*
[2]*Department of Chemistry, Arba Minch University, Arba Minch, Ethiopia*
[3]*Centre for Food Technology, A. C. Tech., Anna University, Chennai, India*

* *Corresponding author:* k.radhakrishnan@cit.ac.in

2.1 Introduction

Packaging is an important process needed to maintain the quality of food products till consumption. It decreases the spoilage of food and leads to effective distribution. Food packaging segment has been growing rapidly through the progresses in material science and technology along with the rising demands from the consumers towards minimally processed foods with enhanced shelf life [1-3]. At present, packaging is not only required for the safe transportation and extending the shelf life of food products, but is also essential to assist the end-use of products and to communicate with the consumers at various stages. Owing to this, the packaging industry has moved to the third position worldwide among other industries and contributes almost 2% of the GNP in developed countries [4,5].

Innovations in food packaging have been boosted by the increased use of packaged foods, rising demand of ready-to-eat or ready-to-serve foods and mounting use of smaller sized food packages [6]. Another major reason for the developments in food packaging is food wastage. The statistical report from FAO indicates that around one-third or over 1.3 billion MT of food products have been wasted every year. The wastage of food begins at the very first stage of food supply chain (harvesting), and it extends up to the last stage (consumption) [7]. If the food wastage is reduced substantially, the rising food demand from the growing world population may automatically be re-

Recent Trends in Nanobiotechnology, edited by Prakash Saudagar and K. Divakar
© 2019 Central West Publishing, Australia

solved. Therefore, a decrease in food wastage by proper food packaging would assist in sustainable food production [8]. The use of novel polymeric materials for packaging would be an effective solution in this respect.

The packaging techniques of today's world perform collectively with a mixture of different processes and agents to protect food from microbial growth and other contaminating factors [9,10]. The problems associated with these can be solved by the use of advanced packaging technologies like modifying the gas composition of packages, controlling the pressure and moisture inside the package, incorporating active component in packages, use of different chilling systems, fermentation, irradiation, etc. [11].

Nanotechnology is a potent multidisciplinary tool for the creation of innovative materials. It has been estimated that by 2020, nanotechnology will contribute $3 trillion to the world economy, producing job opportunities for 6 million people in various industrial sectors [12]. The market value for the activities related to nanotechnology application in food packaging was estimated around US$ 4.13 billion worldwide in 2008, and it was expected to grow at the rate of 12%. In this connection, nanotechnology may offer a key momentum to the development of cutting-edge packaging technologies for fulfilling the expectations of customers [13].

Nanotechnology is an integrative field related to the preparation of structures, devices or materials which possess one or more dimensions at nanoscale (100 nm). With the particle size in the nano-range, the synthesized materials exhibit characteristics that are pointedly distinct from the properties of the materials of normal size [14]. Usually, nanomaterials may be divided into three categories: particulates, platelets and fibers [14,15]. The surface-to-volume ratio of the nanomaterials is very high owing to their size. The properties of nanomaterials can be improved when mixed with suitable polymers. By this process, the mechanical strength, thermal properties and electrical conductivity of the nanomaterials can be enhanced [16]. Therefore, nanomaterials exhibit promising aspects as suitable structures for active and intelligent packaging with better mechanical and barrier properties.

2.1.1 Limitations of Present Packaging Systems and Prospects for Nanotechnology based Packaging Systems

The food packaging sector of the food processing industries is facing numerous challenges related to food packaging materials which include the viability of materials, possibility of recycling, poor mechanical and barrier properties, etc. Though metal and glass containers possess excellent barrier properties, but plastics are preferred owing to the challenges in the bulk transportation of other materials. Plastic materials are chosen due to their unique characteristics like light weight, elastic properties, ease in moldability and low-cost. Therefore, nearly 50% of the plastic materials produced by plastic industries are used for food packaging purposes [17]. However, most of the plastic materials are produced from petroleum sources which are basically non-renewable. Thus, these plastics are non-biodegradable in nature, thus, the disposal of plastics become a great environmental issue.

One of the important problems with present packaging materials is poor barrier performance against water vapor and gases. The fresh food products like fruits and vegetables have to be packed in the materials that have good oxygen transmission rate, whereas the processed foods require packaging materials of different characteristics. Therefore, creating product-specific packaging materials to protect the quality of different types of foods is a difficult task with available sources (thermoplastics). Modifications such as the use of polymer blends and multilayered polymeric materials have been experimented to improve the functional properties of thermoplastics, however, they generally do not produce desirable results due to high production cost and complexity in the recycling of these materials [2].

Accomplishing an appropriate shelf-life of food products is a paramount task for the food processors while keeping the optimal food quality and safety. The growth of microorganisms due to contamination and difficulties in temperature maintenance, nutrient losses are some of the problems to be solved in this process.

2.2 Nanoparticles and Nanocomposites

The food packaging technology, centered on the use of nanomaterials, has been known as nano-food packaging technology, and it has turned out to be a well-established area in the field of nanotechnology. It serves as a substitute to the conventional food packaging methods. Lower quantities of additives called fillers are added to enhance the characteristics of traditional packaging materials [18]. Inclusion of natural nanomaterials as fillers has been observed to be an excellent approach to improve the functional characteristics of the synthetic as well as biodegradable packaging materials. Generally, nanomaterials are formed through two main approaches: top-down and bottom-up. In the first method, the nanomaterials are synthesized by decreasing the size of the bulk materials through physical and chemical processes. In the second method, the nanostructures are created by combining molecules which have the potential of self-assembly [19]. Nanomaterials used for different food packaging applications are presented in Table 2.1.

Table 2.1 Application of nanomaterials in food packaging

Nano-material	Packaging matrix	Level of nano-material (%)	Food item	Packaging function	Ref.
Montmoril-lonite (mmt)	Chi-tosan/acetic acid	5	tryptic soy broth (TSB), and brain heart in-fusions (BHI)	Anti-mi-crobial ac-tivity (AA)	[20]
mmt	Potato starch, po-tato starch/de-gradable polyester	4	Lettuce, spinach	-	[21]
mmt	Cornstarch	1	-	Improved modulus and strength (IM&S)	[22]
mmt	Cornstarch	5	-	IM&S	[22]
mmt	Cornstarch	7	-	IM&S	[22]
laponite	Cornstarch	-	-	IM&S	[22]
Chitosan modified mmt	Cornstarch	-	-	IM&S	[22]
Organically modified mmt	-	5	TSB and BHI	AA	[20]

Ag	-	5	TSB and BHI	AA	[20]
Ag	-	5	TSB and BHI	AA	[20]
Ag zeolite	-	20	TSB and BHI	AA	[20]
Ag	Polyvinylpyr-rolidone	-	Aspara-gus	AA	[23]
Ag, ZnO	Low-density polyethylene (LDPE)	-	Orange juice	AA	[24]
Ag, ZnO	Low-density polyethylene (LDPE)	0.25-1	Orange juice	AA	[25]
Ag	Absorbent pad	-	Poultry meat	AA	[26]
Ag	Cellulose pad	-	Fresh-cut melon	AA	[27]
Ag	Cellulose pad	-	Beef meat ex-udates	AA	[27]
Ag contain-ing polyeth-yleneoxide	Polyethylene	-	Apple juice	AA	[28]
Titanium di-oxide (TiO$_2$)	Polypropyl-ene	-	lettuce	AA	[29]
TiO$_2$	Polyethylene oxide (PEO), poly-ethylene glycol (PEG) and polyvinyl chloride (PVC)	12.5	-	O$_2$ scav-enging	[30]
Ag, TiO$_2$, ka-olin	PE	-	Chinese jujube	AA	[31]
Ag	Absorber	-	Kiwi and melon juices	AA	[32]

Ag, TiO$_2$, kaolin	LDPE	-	Strawberry	AA	[33]
Ag$_2$O	LDPE	-	Apple slice	AA	[34]
ZnO	Polyvinyl chloride (PVC)	-	Fuji apple cuts	AA	[35]
Cu	Cellulose absorber	-	Melon and pineapple juices	AA	[32]
Multi walled carbon nanotube based sensor	Ultrathin polymer substrates	-	-	Food borne pathogens, temperature and moisture level.	[36]
Nanoter™ (organophillic surface modified kaolinite)	PET	5		Improve O$_2$ barrier and decreased water permeability	[37]
Nanoter™ (organophillic surface modified kaolinite)	PET	1		Improve O$_2$ barrier and decreased water permeability	[37]

2.2.1 Clay

The first novel nanocomposite developed for the food packaging purpose is nanoclay assimilated with polymer. Nanoclays have been used more commonly as the nanoparticles in food packaging and nearly 70% of the nanocomposite have been manufactured with nanoclays [38]. The large-scale production of nanocomposite with clay has been attributed to the ability of clay to improve the properties of the packaging materials along with its abundant availability and low cost [39].

Montmorillonite (MMT) is the most extensively used clay for the production of nanocomposites. MMT clays come under the category of 2:1 layered phyllosilicates or smectites, containing aluminum or magnesium hydroxide layer placed between two silicon oxide tetrahedral layers. Thickness (1 nm) and lateral extension (100-500 nm) of MMT platelets lead to high aspect ratio (100-500). The clay structure is generally contained of a large number of silicate sheet layers intermittently arranged as tactoids of 8-10 nm thickness [40].

The nanocomposites are termed as exfoliated if the nanoparticles are completely dispersed in the polymer matrices [41]. As mentioned earlier, nanoclay contains a nanoscaled layer structure, and the nanolayers must be exfoliated within the polymer matrix, so that the high surface area of the nanoclay can be utilized effectively (>750 m^2/g). After the integration of nanoclay in the polymer matrix, the movement of gases through the matrix is prevented, thus, enhancing the barrier properties of the composites. The phenomenon behind the better performance of nanocomposites may be centered on the creation of a maze structure by the clay plates which makes the movements of molecules through the film difficult. The travelling path distance may depend on the characteristics of the clay fillers used in the composites [41]. Thus, nanoclays in the packaging films enhance the shelf life of highly perishable foods by reducing the gas transmission rates. The use of bottles with nanoclays have also been used in brewing industries where the shelf life of beer has been generally extended from 11 weeks to 30 weeks [38].

Apart from the developments in the gas barrier properties of the packaging materials, the nanoclays can also be used to transport the anti-microbial compounds and other additives. Many research studies have stated the ability of nanoclays in stabilizing the additives and effectively regulating the transportation of anti-microbial agents [42,43]. The regulation over the distribution of anti-microbial agents has several advantages in comparison with spraying or dipping of anti-microbial compounds. It is very much essential for the long-time storage of foods as well as to keep the necessary characteristics, such as taste and aroma, of the food systems.

2.2.2 Silver

Silver (Ag) nanoparticles have become a promising substance with several applications, including food packaging. Ag nanoparticles can be synthesized by two methods: *ex situ* and *in situ*. In the *ex situ* method, Ag$^+$ ions are reduced by borohydride and subsequently dispersed into a polymerizable medium. However, regulating the mono-dispersity of the nanoparticles is a challenge. In the second method, the stable silver nanoparticles are formed using precursors in a polymerizable medium, and the method has superior dispersion capacity than the *ex situ* method. During the synthesis of Ag particles from AgNO$_3$, the reduction of silver salt yields silver nanoparticles of various diameters. However, the silver nanoparticles from the physical process are of uniform size with high dispersion quality [44]. The low redox potential of Ag nanoparticles may facilitate smaller Ag particles to discharge more Ag$^+$ in comparison with the corresponding bulk material. In addition, silver nanoparticles can also withstand high temperatures [45].

Ag nanoparticles are very effective against both types of bacteria (Gram-positive and negative) and exhibit high potential against multidrug-resistant microorganisms [46]. Ag nanoparticles have also been distributed uniformly in packaging films, where the particles maintain their size. These films can exhibit anti-bacterial activity for improving the shelf life of foods. This anti-bacterial effect can also be proceeded without Ag nanoparticles where the packaging material acts as a carrier of silver nano-reservoirs [47]. The size, quanta and large surface area increases the interaction with bacteria and consequently improves the effectiveness in comparison with normal Ag particles. It is estimated that the efficiency of Ag nanoparticles is 10-100 times more than AgNO$_3$. Ag nanoparticles have the potential to arrest the growth of microorganisms from the very first incident [48].

Apart from causing direct cell damage, several reports suggested that the anti-microbial activity of Ag nanoparticles may happen through other mechanisms also. The anti-bacterial activity of Ag particles may be stimulated by its contact with dissolved O$_2$. Lok *et al.* [49] stated that the oxidized surface of Ag atoms of Ag nanoparticles is a vital portion to achieve the anti-bacterial effect. The toxicity of Ag nanoparticles has also been observed to depend on the release of reactive oxygen species [50]. Further, metallic Ag nanoparticles can perform as an effective carrier to release of Ag ions after attachment with cell membrane within a short time [51], thus, preventing ATP synthesis and DNA replication. Therefore, the anti-microbial effect is dependent on the availability of Ag ions for acting against bacteria [52], and the toxicity of Ag particles is fully dependent on their size. Maximum anti-bacterial effect is observed with Ag particles with size between 1-10 nm. In addition, the degree of aggregation, ion charge, solubility and surface coating of particles also influence the anti-microbial effect of Ag nanoparticles [12]. Apart from the anti-bacterial effect, Ag nanoparticles also show greater action against ethylene produced by fruits during the ripening process. Therefore, fruit ripening due to the accumulation of ethylene can be prevented and shelf life of fruits can be extended with Ag nanoparticles.

The combination of Ag nanoparticles with zeolites and gold nanoparticles can also be employed. The synergetic effect can be perceived against some microbes with these combinations of nanoparticles used in the packaging materials. The anti-bacterial activity is higher with Ag nanoparticles used in combination with zeolites/gold nanoparticles. However, no reports are observed regarding the commercial use of Ag nanoparticles with zeolites/gold nanoparticles yet [53].

2.2.3 Zinc Oxide (ZnO)

Generally, ZnO nanoparticles are synthesized through physical or mechano-chemical treatments [54]. The manufacturing of ZnO nanoparticles can also be accomplished via chemical synthesis using various precursors as well as thermal and hydrothermal treatments [55]. Engineered ZnO nanoparticles are preferred over Ag nanoparticles due to appearance, UV protective properties and low production cost. In addition, ZnO nanoparticles may also be used to produce cost-effective packaging materials. The inclusion of ZnO nanoparticles in the packaging materials improves effective barrier properties, stability and mechanical strength [55].

The anti-bacterial effect of ZnO nanoparticles is based on the ability of the particles to penetrate (or contact with) the bacteria. The activity of ZnO on bacteria as a bactericidal or bacteriostatic agent is also based on the concentration of the ZnO nanoparticles. It was reported that the ZnO nanoparticles had effective performance against Gram-positive bacteria than the Gram-negative bacteria [56]. The activity of ZnO nanoparticles was observed to be increased by the presence of light, and it also increased as the particle size reduced [57]. The addition of ZnO nanoparticles in the packaging material did not show any adverse reaction between polymer material and food, thus, increasing the shelf life of food materials [55].

2.2.4 Titanium Oxide

Among different production methods, sol-gel processing has been established as a common method for the synthesis of TiO₂ nanoparticles. Commonly, TiO₂ particles have the ability to absorb the light waves of shorter wavelength, thus, TiO₂ particles have been studied widely for their UV blocking. The polymers containing TiO₂ particles may serve as effective inhibitors of oxidation due to UV light blocking and maintaining the transparent properties of the packaging materials at the same time [12]. The photocatalytic properties of TiO₂ particles are helpful to develop the anti-microbial activity under UV A. However, the detrimental effects of TiO₂ nanoparticles have also been reported in the food packaging materials such as EVOH [58] and chitosan. TiO₂ nanoparticles have been observed to display effective anti-bacterial activity against Gram-positive bacteria than Gram-negative bacteria [59]. Further, the TiO₂ nanoparticles can also be used in the active packaging as active O₂ scavengers [30].

2.2.5 Copper Oxide

Cu nanoparticles can be synthesized by thermal or sonochemical processing of copper hydrazine carboxylate complexes in aqueous media. However, Cu nanoparticles are more prone to oxidation because of their low redox potential [32]. The finely distributed Cu nanoparticles are observed to be more efficient sanitizing agents due to their very small size. The smaller size of Cu nanoparticles enables them to produce more Cu ions, thus, permitting them to interact firmly with biological matter [60]. The action of Cu nanoparticles against bacteria can be observed by allowing CuO to react with borojydride in media enriched with ammonia [61].

2.2.6 Carbon Nanotubes

Carbon nanotubes are essentially of cylindrical shape and consist of single-wallled or multi-walled structures with diameter in nanoscale. Carbon nanotubes are observed to have high aspect ratios and large elastic modulus (approx. 1 TPa) [62]. Mainly, carbon nanotubes are incorporated into plastic materials to improve the mechanical properties, however, they also act as potent anti-microbial agents by causing the cell membrane rupture and other irrevocable damage [63].

2.3 Bionanocomposites

The polymeric materials which can be degraded by the action of the naturally occurring organisms (involved in any step in the degradation process) are termed as biopolymers. However, the major disadvantages of biopolymers include poor mechanical and barrier properties, which restrict their use in industrial applications. Especially, the properties like brittleness, poor resistance against heat, high gas and vapor permeability, etc., have restricted their industrial use [64,65].

It has been recommended that the limitations of biopolymers can be overpowered by the application of nanocomposites. Nanocomposites display improved mechanical, heat resistance and barrier properties in comparison with the traditional polymers and composites [66]. For example, montmorillonite clay has been reported to impart effective mechanical and barrier properties to polymers, especially polyamides [67]. It is anticipated that the nanocomposites can resist the stresses during heat treatment, transportation and storage of food [66]. Due to their improved mechanical properties, nanocomposites may also reduce the amount of source materials.

Nowadays, the research on biodegradable polymers has seen increasing trend in order to make them suitable for different applications [68]. Thus, the biodegradable (both natural & synthetic) polymers are formulated with silicate layers for excelling the properties and keeping their biodegradability. Especially, the barriers properties towards preventing the moisture absorbance have been observed to improve as the clay layers prevent the diffusion of moisture into the polymer matrix. Also, nanocomposites can be developed by adding lower quantities of fillers (<5 wt%). For example, poly (butylene succinate) (PBS)/Cloisite 30 B nanocomposites were reported with effective properties [69]. Starch/MMT nanocomposites were formulated by adding different concentrations of filler ranging from 0-11% [70]. As the filler concentration was increased, the tensile strength of nanocomposites increased constantly up to 8% of filler. At the same time, the tensile strain was declined on raising the filler level. Huang *et al.* [71] stated that the incorporation of clay at 5% content improved the tensile strength and strain of starch/MMT nanocomposites up to 450% and 20%, respectively. An increase in the tensile strength (from 8.77 to 15.43 MPa) of soy protein/MMT nanocomposites was also observed by Chen and Zhang, when the MMT level was raised to 16% [72]. Similar findings about the tensile properties of bionanocomposites were reported in other studies [73-75].

Generally, the polymer nanocomposites possess good barrier properties against gases and water vapor. The decrease in the gas permeability of polymer nanocomposites is influenced by the type of clay, aspect ratio of clay particles and structure of polymer matrix. Usually, the polymers with fully exfoliated clay minerals with high aspect ratio show greatest gas barrier properties [76]. Rhim [75] pointed out that the water vapor permeability (WVP) of agar/clay (Cloisite Na+) nanocomposite films reduced exponentially with increasing the filler level from 0 to 20 wt%. Similar findings on the WVP of bionanocomposite generated using other polymers have also reported in other studies [77-79].

In another study, the sensitivity towards water was reduced and the thermo-mechanical properties were enhanced in starch/cellulose nanocomposites [80]. The incorporation of cellulose nanofibers (CNF) also effectively enhanced the barrier properties against water, and WVP reduced from 2.66 to 1.67 g mm/kPa h m². It was also seen that a significant reduction in WVP occurred when at least 10 wt% CNF were added.

Polylactic acid (PLA) can be added with organic clays such as hexadecyl amine-MMT (C16-MMT), dodecyltrimethyl ammonium bromide-MMT (DTA-MMT), Cloisite 25A, etc., for achieving enhanced properties. The oxygen permeability values of the above mentioned composites for the clay addition up to 10 wt% were observed to be less than 50% of the corresponding values of pure PLA, irrespective of organoclay type [81].

The barrier properties of bionanocomposites was reported to be significantly dependent on the aspect ratio of silicate layers, with higher aspect ratio resisting the diffusion of water molecules into the polymer matrix more effectively [76]. The silicate particles present in the polymer matrix make the gas dispersion through the film a convoluted path, thus, enhancing the distance for an effective diffusion. Overall, the improved gas barrier properties of the biopolymer nanocomposites make them suitable for different food packaging applications.

2.4 Food Packaging Developments

2.4.1 Active Packaging

The major function of packaging, i.e. food protection, can be improved by adding active components through active packaging. The active compounds may possess many different functionalities like anti-microbial property, oxygen scavenging, ethylene scavenging, moisture absorption, preservation, etc. By the use of nanotechnology tools, the active compounds can be placed in an active component carrier which can be incorporated in the polymer matrix. The carrier may react with internal (O_2, CO_2, H_2O, microbes, organic compounds, etc.) or external factors (temperature, moisture, etc.) to produce desirable action to improve or maintain the quality or safety of the food products [2]. Some of the active packaging concepts are explained in this section.

Anti-microbials

The potential of anti-microbial systems to inhibit the progression of microorganisms makes them appropriate for food contact applications. The larger surface to volume ratio of the nanoparticles in comparison with micro-particles allow them to interact with more number of microorganisms [82]. Nanomaterials are observed to possess a variety of anti-microbial properties such as prevention of microbial growth [83], bactericidal activities [71] or anti-biotic carriers [84].

Silver nanoparticles (AgNPs) have a wide range of anti-microbial properties against Gram-positive and Gram-negative bacteria, fungi, protozoa and few viruses. Alongside, AgNPs provide additional benefits such as effective thermal stability and low volatility during food processing operations [85]. Due to the high surface to volume ratio, AgNPs are efficient anti-microbial agents than larger silver particles [23]. For confirming this, Damm *et al.* [86] also evaluated the effectiveness of polyamide 6/silver (PA6/Ag) micro-composite comprising 1.9 wt% Ag and a PA6/AgNP nanocomposite having 0.06 wt% Ag. Both samples were inoculated with *E. coli* cultures and kept at room temperature under continuous shaking for 24 h. PA6/AgNP nanocomposite effectively eliminated all *E. coli* even with lower silver content, whereas the micro-composite containing higher amount of silver particles eliminated only 80% of *E. coli* organisms. At the same time, the cytotoxicity and diffusion of AgNPs into human cells may also increase because of their small size [87].

The action of AgNPs on microbial cells may be proceeded through three main mechanisms, although their anti-microbial activity has not been completely explored: hindering ATP synthesis and DNA replication through steady discharge of Ag^+ ions, disruption of cell membranes by AgNPs and production of reactive oxygen species (ROS) by AgNPs and Ag^+ ions [88]. The packaging materials containing AgNPs show anti-microbial activities against all types of microorganisms, particularly bacteria. Sanpui *et al.* [89] evaluated the anti-microbial activity of chitosan-AgNP nanocomposite against *E. coli*. The authors observed that the AgNPs, at very low concentration (2.15%, w/w), considerably reduced the *E.coli* population. The effectiveness of AgNPs was studied by keeping the discs of plain chitosan lactate (CL) and CL films containing AgNPs in a nutrient broth (10^9 ufc/ml). Discs having AgNPs completely inhibited the growth of *E. coli* in the nutrient broth [90].

Tankhiwale and Bajpai [91] prepared filter papers containing AgNPs and tested their activity against *E. coli* by agar diffusion method. A clear zone of inhibition was detected around the disc containing AgNPs, whereas clear microbial growth was noticed around the disc without AgNPs. In another study, AgNPs were incorporated into hydroxypropylmethylcellulose (HPMC) films through continuous stirring [92]. Here, AgNPs of two different diameters (41 and 100 nm) were selected and the effectiveness of HPMC-AgNPs films was evaluated against *E. coli* or *S. aureus*. Discs of HPMC-AgNPs films were kept over the agar medium showed anti-microbial effect against both bacteria. In particular, HPMC discs having AgNPs of 41 nm produced better activity than the discs having AgNPs of 100 nm.

Metal oxide nanoparticles are well suited for different food packaging applications. Nanoparticles from metal oxides like titanium dioxide (TiO_2), zinc oxide (ZnO) and magnesium oxide (MgO) have been well known for their anti-bacterial effect due to the production of ROS [56]. The preference of metal oxide nanoparticles over organic materials is also due to their higher stability. Titanium oxide nanoparticles have been largely used as surface coatings of polymers owing to their disinfecting abilities. Per-oxidation of phospholipids in the microbial cell membrane has been observed to be enhanced by the titanium oxide mediated photocatalysis which results in the damage of cell membrane. Zinc oxide nanoparticles have been observed to be effective against Gram-positive than Gram- negative bacteria [56], and the mechanism of action is mediated by the production of ROS partially. The anti-bacterial activity of ZnO nanoparticles against some bacterial strains like *L. monocytogenes*, *S. enteritidis* and *E. coli* has also been presented by Jin *et al.* [93]. From the study, it can be observed that the anti-microbial action of ZnO nanoparticles greatly depends on their concentration in the packaging materials. Also, the anti-bacterial effect of ZnO nanoparticles may be bacteriostatic or bactericidal. In the same study, polystyrene (PS) films were prepared with ZnO nanoparticles and analyzed for their anti-bacterial activity. The,

PS-ZnO films, however, did not exhibit any anti-bacterial effect which may reduce the potential of their use in packaging applications.

Al-Hazmi *et al.* [94] prepared magnesium oxide nanowires of 6 nm diameter and evaluated their anti-bacterial effect against *E. coli* and *Bacillus sp.* From the results, it was observed that the action of MgO particles against the bacterial strains was solely dependent on their concentration. At higher concentrations, the release of superoxide anions disrupts the bacterial cell walls.

Oxygen and Ethylene Scavengers

As oxygen enters the food package, it may decrease the shelf life of the food as it leads to numerous degradation reactions (rancidity, browning, growth of aerobic microorganisms, degradation of vitamins and flavor compounds, etc.). The key element behind these issues is the amount of oxygen existing inside the food package. By removing or decreasing oxygen to safer levels, the product quality as well as the shelf life can be improved. Another major issue with the fruits and vegetable packaging is the generation of unwanted compounds which decrease the shelf life. Especially, the production of ethylene gas by climacteric fruits during storage may hasten the ripening process and reduce the product storage life. So, the use of nanostructures in food packaging may provide a solution to these challenges [2].

One of the most commonly used polymer categories for the food packaging applications is polyolefins. Polyolefins display better sealing and barrier properties among the other polymer materials. The polymers from this group show poor resistance towards oxygen diffusion. Thus, the use of these polymers for oxygen sensitive materials is limited. Modification of high density polyethylene (HDPE) films by adding kaolinite particles may produce polymer films with oxygen scavenging ability [95]. The alteration in the polymer matrix exhibited a significant effect on the oxygen scavenging activity. The oxygen scavenging characteristic of the film was mainly dependent on the oxygen trapping capacity and elongated convoluted diffusion pathway.

One of the metal oxides generally used in a varieties of packaging products is TiO_2. TiO_2 nanoparticles can be photo-induced by the application of UV rays (<388 nm wavelength) which leads to the charge segregation as the electrons move from the valence to the conduction band. Consequently, the electrons are deposited on the external area of TiO_2. The main driving force of this reaction is the rate at which the electrons are transferred to oxygen present in the atmospheric air. Various studies have been conducted for determining the oxygen scavenging capacity of TiO_2. Li *et al.* [96] developed glass and acetate films incorporated with nanocrystalline TiO_2 and confirmed the ability of nanocrystalline TiO_2 in creating the deoxygenated environment inside the package. The authors stated that the oxygen scavenging reaction followed first order reaction kinetics at the rate of 70 s^{-1}. TiO_2 may also be used for ethylene scavenging in fruits and vegetable packaging due to its photo-catalytic ability for slowing down the ripening process in climacteric fruits. The traditional ethylene scavengers have a major drawback of restricted scavenging ability, i.e. the scavenging ability of the materials is completely dependent on the concentration in the packaging materials. However, TiO_2 based packaging materials have unrestricted scavenging ability as TiO_2 is not devoured during the scavenging process. Maneerat and Hayata [97] prepared polypropylene (PP) films with TiO_2 for eliminating the ethylene gas produced during the packaging of fruits. The authors also studied the efficiency of the particles of different sizes ($\sim$ 5 µm and $\sim$7 nm). From the results, it was clear that the PP films with nanoparticles had better scavenging ability over the micro-particles incorporated PP films.

Ye *et al.* [98] demonstrated the photo-electrocatalytic based ethylene scavenging of TiO_2 incorporated on activated carbon felts. Though the packaging material possessed sufficient ethylene scavenging capacity when treated with UV, the authors noticed that the efficiency was increased when a bias voltage was imposed. The concept of using UV radiation along with applied potential may decrease the ripening rate of climacteric fruits for longer storage conditions by eliminating ethylene vapor continuously.

2.4.2 Intelligent Packaging

The concept of intelligent packaging revolves around an "ON/OFF" switching mechanism in the package with respect to a change in an internal/external factor and informing the product's condition to the customers or end users [99]. The packaging system is developed by integrating an external discrete component in the final package. Three key strategies are used to develop the intelligent packaging systems: (i) indicators, which convey the status of food materials to the consumers, (ii) data carriers, like barcodes and radiofrequency identification tags (RFID), which are mainly employed to monitor the product quality throughout the supply chain and (iii) sensors, which permit the quick analysis of the compounds in the food products [100].

Indicators

The key objective of employing indicators in the packaging system is to inform the consumer about the presence/absence of a specific compound or the level of a particular constituent or a class of compounds. Generally, the message regarding the product quality is conveyed to the consumers by prompt visual change, e.g. use of various colors [101]. The type of information passed to the consumers through the indicators may be of qualitative or semi-quantitative in nature. The indicators used for different food packaging applications can be classified in three categories: time-temperature indicators, freshness indicators and gas indicators [102].

Temperature indicators can be classified into two types: simple temperature indicators and time-temperature integrators (TTIs). Temperature indicators state the consumers about the temperature at which the food product has been heated or cooled with respect to the reference temperature. These also alert the consumers about the possibility of microbial spoilage and changes in the protein quality during low temperature processing conditions. TTIs are grouped under the category of first generation indicators which are used to display any adverse changes in the product temperature with respect to time. The working principle of these indicators centers on the mechanical, chemical, electrochemical, enzymatic or microbiological changes, generally conveyed by an apparent color change or color development [103]. TTIs have been well accepted because of their role in determining the food spoilage as the functioning of TTIs are based on time-temperature relationship.

Freshness indicators are used to check the changes in the food quality during the supply chain. The product freshness may be degraded due to the exposure to the adverse environment or exceeded shelf life. The data concerning the product quality with respect to biological or chemical changes has been reported to be conveyed directly by freshness indicators to the consumers. Especially, the quality of sea food can be monitored by freshness indicators, and the working mechanism is based on the production of volatile basic nitrogen content formed due to food spoilage. The quality of meat products can be evaluated by using hydrogen sulfide indicators. During ageing of meat, H_2S is liberated, and it can be associated with the meat pigment, myoglobin, which has been accepted as a quality attribute of the meat product. Based on the above mentioned principle, Smolander et al. [104] formulated a package with a freshness indicator to determine the poultry meat quality stored in modified atmospheric conditions. Other indicators have also been developed, depending on the affinity to the microbial metabolites like, ethanol, diacetyl and CO_2 [105]. Some of the commercially available freshness indicators are Toxinguard® by Toxin Alert Inc., to monitor *Pseudomonas sp.* growth and SensorQ™ by FQSI Inc., for monitoring the quality of different meat-based food products [101].

Gas indicators are used to detect the changes in the gas composition inside the package. Gas composition may change due to the absorption of gases by packages, growth of microorganisms, chemical and enzymatic processes [99]. Gas indicators may also be used to evaluate the efficiency of various elements (CO_2 and O_2 scavengers) used in the active packaging as well as to know about the leakages in the package.

Data Carriers

Data carriers are involved to ensure the safety and quality of the food products by facilitating the information throughout the supply chain. Data carriers do not provide any information regarding the product quality directly, instead they are used to provide the information related to automatization, traceability, theft prevention, etc. [106]. Mainly, the data carrier devices employed in food industries are classified into two categories: barcodes and RFID tags.

Barcodes have been used widely in the food supply chain as they are available at low cost as well as easy to use [107]. A barcode consists of space and bars which are used to denote 12 digits of data. The data stored in the barcode can be identified by a scanner and is used to transmit the data to a system where it is stored and processed [108]. Initially, barcodes of one dimensional (1-D) arrays were created. The functioning was very similar to the working of a laser beam used to cut a horizontal slice from the vertical code bars. When the beam moved over the symbols, the time taken by the scanner to scan the dark bars and spaces was calculated. By using a lookup table, the individual characters were decoded by relating the time taken by the scanners during the scanning process. The major drawback of 1-D barcodes is the restricted storage capacity [109].

To overcome the problem of limited storage, two-dimensional (2-D) barcodes were developed. Here, the information is stored with the help of dots and spaces which is organized in a form of an array or a matrix. It permits the storage of larger amount of data in a confined space. For example, Portable Data File (PDF) 417, a 2-D symbol that can be used to store 1.1 kB of data can be kept within the space of a UPC barcode [110]. Another type of 2-D barcode, which is known as Quick Response (QR) 2-D barcode, can be used to store more amount of data in comparison with normal 2-D barcodes. Specialized scanning device, which have the capacity to scan the symbols both vertically and horizontally (two dimensions), is required to decode the 2-D barcodes [111].

Another class of more advanced data carrier devices is RFID tags. RFID tag consists of three major parts: (i) a tag, (ii) a reader and (iii) middleware [112]. The distinctive characteristic of RFID tags is the larger number of codes that can be stored, and these can be transferred over a long distance. Thus, the automatic product verifications and product traceability can be enhanced. Generally, RFID tags have been used in the food processing industries to monitor the cold chain operations [113], analyze meat and meat products [114], estimate product shelf life [115] and achieve product identification and traceability [116].

Sensors

One of the most promising technology for packaging applications in near future is the use of sensors in packages [117]. Essentially, sensor may be a device or system, containing control and processing electronics, an interconnection network and software [118]. Sensor can be employed to evaluate, detect or quantity energy or matter, by producing a signal for the detection or quantification [119]. Generally, sensors consists of four major parts: (i) a receptor, which functions as a sensing portion, denoted by a sampling area where the surface chemistry occurs, (ii) a transduction element, used to quantifying part of the sensor, (iii) signal processing electronics and (iv) a display unit [120].

In the past few years, sensors of different types have been developed for various food packaging functions, like electrochemical sensors [121] and luminescence sensors [122]. Electrochemical sensors are categorized under chemical sensors, where an electrode is employed as the transduction element. In luminescence sensors, the quantification of signals is carried out once the analyte is placed and consecutively production of solid-phase luminescence (SPL) or to its equivalent solid-matrix luminescence (SML) occurs. The expressions are correlated with the level of analyte present in the sample under suitable conditions [123]. Another two important classes of sensors under the intelligent packaging systems are biosensors and gas sensors.

2.5 Regulatory Issues

At present, the use of nanotechnology advancements in the food processing sector has been increasing at a faster rate. The accomplishment of these applications is strongly reliant on the consideration of regulatory issues. Unfortunately, no clear or defined regulations exist for the use of nanomaterials in the industries. Among different countries, European countries have tried to formulate the stringent and defined regulations for the use of nano-products. The existing food laws for the common food products have been followed for the broad spectrum of nano-based food products.

Recently, the European Parliament has changed its stand towards the use of nanomaterials in food applications and has supported the European regulatory debate. In 2009, it was made mandatory that the commission reviewed the legislation within two years to confirm that the judicial provisions and instruments of execution show the required features of nanomaterials to which the labor, consumers and the environment may be exposed [124]. Subsequently, the essential guidelines specifying the requirements of nanomaterials have been enabled as laws/regulations. However, these do not contain regulations specifically for each category/type of food. Thus, it is very difficult for the food industries to have a clear idea on the relevant regulatory framework [125].

On the whole, a few guidelines specifying the requirements of nanomaterials have been developed. Apart from these guidelines, the discussion on nano-regulation in EU has taken place over a long time period and has resulted in the formation of EU laws concerning the regulation over nanomaterials. It can be observed that the term soft law is used to specify EU measures that include communications, resolutions, endorsements, views, rules and procedures. On the contrary, the term hard law is used to point out EU activities on directives, regulations and decisions. Soft laws are also termed as private laws, framed by the stakeholders with support from private sector and civil body representatives, all involved in the decision making processes.

2.6 Conclusion

Nowadays, the application of nanotechnology in food industry can be seen in most of the processing steps. Nanotechnology has provided satisfactory results in food packaging as well as in food safety. The mechanical and barrier properties of the packaging materials have been enhanced by the addition of nanomaterials, thereby, resulting in reduced use of raw materials and lesser waste generation. A major issue in the food processing industries is the need of longer time periods for analyzing the quality of food products. However, the nanotechnology developments have reduced the time required for sample preparations and analysis. With the help of intelligent packaging technology, various food contaminants can be detected instantaneously which is considered as a most promising development. However, the potential hazards to human health and environment are not defined well. It can be mentioned that the nanoparticles should only be used in food applications once the safety of the consumers has been assured by a set of vigorous tests, as given in the report published by the Institute of Food Science and Technologists in 2006. It also specifies that special attention should be given to the consumer queries regarding the use of nanotechnology in food processing. The governments should, thus, formulate the regulations and proper labelling mechanisms in order to increase the consumer acceptability.

References

1. Han, J. H. (2005) New technologies in food packaging: Overview. In: *Innovations in Food Packaging*, Han, J. H. (ed.), Elsevier Academic Press, The Netherlands.
2. Mihindukulasuriya, S. D. F., and Lim, L. T. (2014) Nanotechnology development in food packaging: A review. *Trends in Food Science and Technology*, **40**(2), 149-167.
3. Dobrucka, R., and Cierpiszewski, R. (2014) Active and intelligent packaging food-research and development-a review. *Polish Journal of Food and Nutrition Sciences*, **64**(1), 7-15.
4. Han, J. H. (2005) *Innovations in Food Packaging*, 2nd edition, Academic Press, USA.
5. Robertson, G. L. (2005) *Food packaging: Principles and Practice.* CRC

Press, USA.

6. Restuccia, D., Spizzirri, U. G., Parisi, O. I., Cirillo, G., Curcio, M., Lemma, F., Puoci, F., Vinci, G., and Picci, N. (2010) New EU regulation aspects and global market of active and intelligent packaging for food industry applications. *Food Control*, **21**, 1425- 1435.

7. *WHO Model Lists of Essential Medicines* (2017). Online: http://www.who.int/medicines/publications/essentialmedicines/en/index.html [accessed 19th May 2019].

8. Morris, M. A., Padmanabhan, S. C., Cruz-Romero, M. C., Cummins, E., and Kerry, J. P. (2017) Development of active, nanoparticle, antimicrobial technologies for muscle-based packaging applications. *Meat Science*, **132**, 163-178.

9. Cruz-Romero, M. C., and Kerry, J. P. (2011) Packaging of cooked meats and muscle based, convenience-style processed foods. In: *Processed Meats*, Kerry, J. P., and Kerry, J. F. (eds.), Woodhead Publishing, USa, pp. 666-705.

10. Rodríguez-Calleja, J. M., Cruz-Romero, M. C., O'Sullivan, M. G., García-López, M. L., and Kerry, J. P. (2012) High-pressure-based hurdle strategy to extend the shelf life of fresh chicken breast fillets. *Food Control*, **25**(2), 516-524.

11. Chouliara, E., Badeka, A., Savvaidis, I., and Kontominas, M. G. (2008) Combined effect of irradiation and modified atmosphere packaging on shelf life extension of chicken breast meat: Microbiological, chemical and sensory changes. *European Food Research and Technology*, **226**(4), 877-888.

12. Duncan, T. V. (2011) Applications of nanotechnology in food packaging and food safety: Barrier materials, antimicrobials and sensors. *Journal of Colloid and Interface Science*, **363**, 1-24.

13. Youssef, A. M., and El-Sayed, S. M. (2018) Bionanocomposites materials for food packaging applications: Concepts and future outlook. *Carbohydrate Polymers*, **193**, 19-27.

14. Youssef, A. M., Bujdos, T., Hornok, V., Papp, S., Kiss, B., Abd El-Hakim, A., and Dékány, I. (2013) Structural and thermal properties of polystyrene nanocomposites containing hydrophilic and hydrophobic layered double hydroxide. *Applied Clay Science*, **77-78**, 46-51.

15. Youssef, A. M., Kamel, S., and El-Samahy, M. A. (2013) Morphological andantibacterial properties of modified paper by PS nanocomposites for packaging applications. *Carbohydrate Polymers*, **98**, 1166-1172.

16. Uskokovic, V. (2007) Nanotechnologies: What we do not know. *Technology in Society*, **29**, 43-61.

17. Rhim, J.-W., Park, H.-M., and Ha, C.-S. (2013) Bio-nanocomposites for food packaging applications. *Progress in Polymer Science*, 1629-1652.

18. Cushen, M., Kerry, J., Morris, M., Cruz-Romero, M., and Cummins, E.

(2014) Silver migration from nanosilver and a commercially available zeolite filler polyethylene composites to food simulants. *Food Additives and Contaminants: Part A*, **31**(6), 1132-1140.

19. Huang, J. Y., Li, X., and Zhou, W. (2015) Safety assessment of nanocomposite for food packaging application. *Trends in Food Science and Technology*, **45**(2), 187-199.

20. Rhim, J.-W., Hong, S.-I., Park, H.-M., and Ng, P. K. W. (2006) Preparation and characterization of chitosan-based nanocomposite films with antimicrobial activity. *Journal of Agricultural and Food Chemistry*, **54**(16), 5814-5822.

21. Avella, M., De Vlieger, J. J., Errico, M. E., Fischer, S., Vacca, P., and Volpe, M. G. (2005) Biodegradable starch/clay nanocomposite films for food packaging applications. *Food Chemistry*, **93**(3), 467-474.

22. Chung, Y.-L., Ansari, S., Estevez, L., Hayrapetyan, S., Giannelis, E. P., and Lai, H.-M. (2010) Preparation and properties of biodegradable starch-clay nanocomposites. Carbohydrate *Polymers*, **79**(2), 391-396.

23. An, J., Zhang, M., Wang, S., and Tang, J. (2008) Physical, chemical and microbiological changes in stored green asparagus spears as affected by coating of silver nanoparticles-PVP. *LWT - Food Science and Technology*, **41**(6), 1100-1107.

24. Emamifar, A., Kadivar, M., Shahedi, M., and Soleimanian-Zad, S. (2010) Evaluation of nanocomposite packaging containing Ag and ZnO on shelf life of fresh orange juice. *Innovative Food Science and Emerging Technologies*, **11**(4), 742-748.

25. Emamifar, A., Kadivar, M., Shahedi, M., and Soleimanian-Zad, S. (2011) Effect of nanocomposite packaging containing Ag and ZnO on inactivation of Lactobacillus plantarum in orange juice. *Food Control*, **22**(3-4), 408-413.

26. Fernandez, A., Soriano, E., Lopez-Carballo, G., Picouet, P., Lloret, E., Gavara, R., and Hernández-Muñoza, P. (2009) Preservation of aseptic conditions in absorbent pads by using silver nanotechnology. *Food Research International*, **42**(8), 1105-1112.

27. Fernandez, A., Picouet, P., and Lloret, E. (2010) Cellulose-silver nanoparticle hybrid materials to control spoilage-related microflora in absorbent pads located in trays of fresh-cut melon. *International Journal of Food Microbiology*, **142**(1-2), 222-228.

28. Del Nobile, M. A., Cannarsi, M., Altieri, C., Sinigaglia, M., Favia, P., Iacoviello, G., et al. (2004) Effect of Ag-containing nanocomposite active packaging system on survival of alicyclobacillus acidoterrestris. *Journal of Food Science*, **69**(8), E379-E383.

29. Chawengkijwanich, C., and Hayata, Y. (2008) Development of TiO_2 powder-coated food packaging film and its ability to inactivate Escherichia coli in vitro and in actual tests. *International Journal of Food Microbiology*, **123**(3), 288-292.

30. Xiao-e, L., Green, A. N. M., Haque, S. A., Mills, A., and Durrant, J. R. (2004) Light-driven oxygen scavenging by titania/polymer nano-composite films. *Journal of Photochemistry and Photobiology A: Chemistry*, **162**, 253-259.

31. Li, H., Li, F., Wang, L., Sheng, J., Xin, Z., Zhao, L., et al. (2009) Effect of nano-packing on preservation quality of chinese jujube (Ziziphus jujuba Mill. var. inermis (Bunge) Rehd) *Food Chemistry*, **114**(2), 547-552.

32. Llorens, A., Lloret, E., Picouet, P., and Fernandez, A. (2012) Study of the antifungal potential of novel cellulose/copper composites as ab-sorbent materials for fruit juices. *International Journal of Food Mi-crobiology*, **158**(2), 113-119.

33. Yang, F. M., Li, H. M., Li, F., Xin, Z. H., Zhao, L. Y., Zheng, Y. H., and Hu, Q. H. (2010) Effect of nano-packing on preservation quality of fresh strawberry (Fragaria ananassa Duch. Cv fengxiang) during storage at 4°C. *Journal of Food Science*, **75**(3), C236-C240.

34. Zhou, L., Lv, S., He, G., He, Q., and Shi, B. I. (2011) Effect of PE/AG2O nano-packaging on the quality of apple slices. *Journal of Food Qual-ity*, **34**(3), 171-176.

35. Li, X. H., Li, W. L., Xing, Y. G., Jiang, Y. H., Ding, Y. L., and Zhang, P. P. (2010) Effects of nano-ZnO power-coated PVC film on the physio-logical properties and microbiological changes of fresh-cut) Fuji" apple. *Advanced Materials Research*, **152-153**, 450-453.

36. Nachay, K. (2007) Analyzing nanotechnology. *Food Technology*, **61**, 34-36.

37. Sanchez-Garcia, M. D., Gimenez, E., and Lagaron, J. M. (2007) Novel PET nanocomposites of interest in food packaging applications and comparative barrier performance with biopolyester nanocompo-sites. *Journal of Plastic Film and Sheeting*, **23**(2), 133-148.

38. Silvestre, C., Duraccio, D., and Cimmino, S. (2011) Food packaging based on polymer nanomaterials. *Progress in Polymer Science*, **36**(12), 1766-1782.

39. Sorrentino, A., Gorrasi, G., and Vittoria, V. (2007) Potential perspec-tives of bionanocompositesfor food packaging applications. *Trends in Food Science and Technology*, **18**(2), 84-95.

40. Rodriguez, F., Sepulveda, H. M., Bruna, J., Guarda, A., and Galotto, M. J. (2013) Development of cellulose eco-nanocomposites with anti-microbial properties oriented for food packaging. *Packaging Tech-nology and Science*, **26**(3), 149-160.

41. De Abreu, D. A. P., Cruz, J. M., Angulo, I., and Losada, P. P. (2010) Mass transport studies of different additives in polyamide and ex-foliated nanocomposite polyamide films for food industry. *Packag-ing Technology and Science*, **23**(2), 59-68.

42. Chakraborti, M., Jackson, J. K., Plackett, D., Gilchrist, S. E., and Burt, H. M. (2012) The application of layered double hydroxide clay

(LDH)-poly(lactide-co-glycolic acid) (PLGA) film composites for the controlled release of antibiotics. *Journal of Materials Science, Materials in Medicine*, **23**(7), 1705-1713.

43. Girdthep, S., Worajittiphon, P., Molloy, R., Lumyong, S., Leejarkpai, T., and Punyodom, W. (2014) Biodegradable nanocomposite blown films based on poly(lactic acid) containing silver-loaded kaolinite: a route to controlling moisture barrier property and silver ion release with a prediction of extended shelf life of dried longan. *Polymer*, **55**(26), 6776-6788.

44. De Azeredo, H. M. C. (2013) Antimicrobial nanostructures in food packaging. *Trends in Food Science and Technology*, **30**(1), 56-69.

45. Cushen, M., Kerry, J., Morris, M., Cruz-Romero, M., and Cummins, E. (2013) Migration and exposure assessment of silver from a PVC nanocomposite. *Food Chemistry*, **139**, 389-397.

46. Li, W. R., Xie, X. B., Shi, Q. S., Duan, S. S., Ouyang, Y. S., and Chen, Y. B. (2011) Antibacterial effect of silver nanoparticles on Staphylococcus aureus. *Biometals*, **24**(1), 135-141.

47. Llorens, A., Lloret, E., Picouet, P. A., Trbojevich, R., and Fernandez, A. (2012b) Metallic based micro and nanocomposites in food contact materials and active food packaging. *Trends in Food Science and Technology*, **24**(1), 19-29.

48. Lara, H. H., Ayala-Nunez, N. V., Turrent, L. D. I., and Padilla, C. R. (2010) Bactericidal effect of silver nanoparticles against multidrug-resistant bacteria. *World Journal of Microbiology and Biotechnology*, **26**, 615-621.

49. Lok, C., Ho, C. M., Chen, R., He, Q. Y., Yu, W. Y., Sun, H., Tam, P. K.-H., Chiu, J.-F., and Che, C.-M. (2007) Silver nanoparticles: partial oxidation and antibacterial activities. *Journal of Biological Inorganic Chemistry*, **12**, 527-534.

50. Kim, J. S., Kuk, E., Yu, K. N., Kim, J. H., Park, S. J., Lee, H. J., Kim, S. H., Park, Y. K., Park, Y. H., Hwang, C.-Y., Kim, Y.-K., Lee, Y.-S., Jeong, D. H., and Cho, M.-H. (2007) Antimicrobial effects of silver nanoparticles. *Nanomedicine: Nanotechnology, Biology and Medicine*, **3**, 95-101.

51. Limbach, L. K., Wick, P., Manser, P., Grass, R. N., Bruinink, A., and Stark, W. J. (2007) Exposure of engineered nanoparticles to human lung epithelial cells: influence of chemical composition and catalytic activity on oxidative stress. *Environmental Science and Technology*, **41**(11), 4158-4163.

52. Magana, S. M., Quintana, P., Aguilar, D. H., Toledo, J. A., Angeles-Chavez, C., Cortaes, M. A., Leóna, L., Freile-Pelegrína,Y., Lópezcd, T., and TorresSánchez, R. M. (2008) Antibacterial activity of montmorillonites modified with silver. *Journal of Molecular Catalysis A: Chemical*, **281**, 192-199.

53. Mbhele, Z. H., Salemane, M. G., Van Sittert, C. G. C. E., Nedeljković, J. M., Djoković, V., and Luyt, A. S. (2003) Fabrication and characteriza-

tion of silver– polyvinyl alcohol nanocomposites. *Chemistry of Materials*, **15**(26), 5019-5024.

54. Casey, P. (2006) Nanoparticle technologies and applications. In: *Nanostructure Control of Materials*, Hannink, R. H. J., and Hill, A. J. (eds.), Woodhead Publishing Limited, UK, pp. 1-31.

55. Espitia, P., Soares, N. d., Coimbra, J. d., de Andrade, N., Cruz, R., and Medeiros, E. (2012) Zinc oxide nanoparticles: synthesis, antimicrobial activity and food packaging applications. *Food and Bioprocess Technology*, **5**(5), 1447-1464.

56. Premanathan, M., Karthikeyan, K., Jeyasubramanian, K., and Manivannan, G. (2011) Selective toxicity of ZnO nanoparticles toward gram-positive bacteria and cancer cells by apoptosis through lipid peroxidation. *Nanomedicine*, **7**(2), 184-192.

57. Yamamoto, O. (2001) Influence of particle size on the antibacterial activity of zinc oxide. *International Journal of Inorganic Materials*, **3**, 643-646.

58. Cerrada, M. L., Serrano, C., Sanchez-Chaves, M., Fernandez-Garcia, M., Fernandez-Martin, F., de Andres, A., et al. (2008) Self-sterilized EVOH-TiO_2 nanocomposites: interface effects on biocidal properties. *Advanced Functional Materials*, **18**, 1949-1960.

59. Xing, Y., Li, X., Zhang, L., Xu, Q., Che, Z., Li, W., et al. (2012) Effect of TiO2 nanoparticles on the antibacterial and physical properties of polyethylene-based film. *Progress in Organic Coatings*, **73**, 219-224.

60. Conte, A., Longano, D., Costa, C., Ditaranto, N., Ancona, A., Cioffi, N., Scrocco, C., Sabbatini, L., Contòd, F., and Del Nobile, M.A. (2013) A novel preservation technique applied to fiordilatte cheese. *Innovative Food Science and Emerging Technologies*, **19**(0), 158-165.

61. Kotelnikova, N., Vainio, U., Pirkkalainen, K., and Seriman, R. (2007) Novel approaches to metallization of cellulose by reduction of cellulose-incorporated copper and nickel ions. *Macromolecular Symposia*, **254**, 74-79.

62. Lau, K. T., and Hui, D. (2002) The revolutionary creation of new advanced materials- carbon nanotube composites. *Composites: Part B*, **33**(4), 263-277.

63. Kang, S., Pinault, M., Pfefferle, L. D., and Elimelech, M. (2007) Single-walled carbon nanotubes exhibit strong antimicrobial activity. *Langmuir*, **23**, 8670-8673.

64. Koh, H. C., Park, J. S., Jeong, M. A., Hwang, H. Y., Hong, Y. T., Ha, S. Y., and Nam, S. Y. (2008) Preparation and gas permeation properties of biodegradable polymer/layered silicate nanocomposite membranes. *Desalination*, **233**(1-3), 201-209.

65. Sorrentino, A., Tortora, M., and Vittoria, V. (2006) Diffusion behavior in polymer-clay nanocomposites. *Journal of Polymer Science, Part B: Polymer Physics*, **44**(2), 265-274.

66. Ray, S., Quek, S. Y., Easteal, A., and Chen, X. D. (2006) The potential

use of polymer-clay nanocomposites in food packaging. *International Journal of Food Engineering*, **2**(4), doi: 10.2202/1556-3758.1149.

67. Cho, J. W., and Paul, D. R. (2001) Nylon 6 nanocomposites by melt compounding. *Polymer*, **42**(3), 1083-1094.

68. Ray, S. S., and Bousmina, M. (2005) Biodegradable polymers and their layered silicate nanocomposites: in greening the 21st century materials world. *Progress in Materials Science*, **50**(8), 962-1079.

69. Lee, S. R., Park, H. M., Lim, H., Kang, T., Li, X., Cho, W. J., and Ha, C. S. (2002) Microstructure, tensile properties, and biodegradability of aliphatic polyester/clay nanocomposites. *Polymer*, **43**(8), 2495-2500.

70. Huang, M., and Yu, J. (2006) Structure and properties of thermoplastic corn starch/montmorillonite biodegradable composites. *Journal of Applied Polymer Science*, **99**(1), 170-176.

71. Huang, L., Li, D. Q., Lin, Y. J., Wei, M., Evans, D. G., and Duan, X. (2005) Controllable preparation of nano-MgO and investigation of its bactericidal properties. *Journal of Inorganic Biochemistry*, **99**, 986-993.

72. Chen, P., and Zhang, L. (2006) Interaction and properties of highly exfoliated soy protein/montmorillonite nanocomposites. *Biomacromolecules*, **7**(6), 1700-1706.

73. Rao, Y. (2007) Gelatin-clay nanocomposites of improved properties. *Polymer*, **48**(18), 5369-5375.

74. Tang, X. Z., Kumar, P., Alavi, S., and Sandeep, K. P. (2012) Recent advances in biopolymers and biopolymer-based nanocomposites for food packaging materials. *Critical Reviews in Food Science and Nutrition*, **52**(5), 426-442.

75. Rhim, J. W. (2011) Effect of clay contents on mechanical and water vapor barrier properties of agar-based nanocomposite films. *Carbohydrate Polymers*, **86**(2), 691-699.

76. Choudalakis, G., and Gotsis, A. D. (2009) Permeability of polymer/clay nanocomposites: A review. *European Polymer Journal*, **45**(4), 967-984.

77. Sothornvit, R., Hong, S. I., An, D. J., and Rhim, J. W. (2010) Effect of clay content on the physical and antimicrobial properties of whey protein isolate/organo-clay composite films. *LWT - Food Science and Technology*, **43**(2), 279-284.

78. Kumar, P., Sandeep, K. P., Alavi, S., Truong, V. D., and Gorga, R. E. (2010) Preparation and characterization of bio-nanocomposite films based on soy protein isolate and montmorillonite using melt extrusion. *Journal of Food Engineering*, **100**(3), 480-489.

79. Tunc, S., Angellier, H., Cahyana, Y., Chalier, P., Gontard, N., and Gastaldi, E. (2007) Functional properties of wheat gluten/montmorillonite nanocomposite films processed by casting. *Journal Of Membrane Science*, **289**(1-2), 159-168.

80. Dufresne, A., Dupeyre, D., and Vignon, M. R. (2000) Cellulose micro-fibrils from potato tuber cells: processing and characterization of starch-cellulose microfibril composites. *Journal of Applied Polymer Science*, **76**(14), 2080-2092.
81. Rhim, J. W., and Ng, P. K. (2007) Natural biopolymer-based nano-composite films for packaging applications. *Critical Reviews In Food Science And Nutrition*, **47**(4), 411-433.
82. Luo, P. G., and Stutzenberger, F.J. (2008) Nanotechnology in the de-tection and control of microorganisms. *Advances in Applied Microbi-ology*, **63**, 145-181.
83. Cioffi, N., Torsi, L., Ditaranto, N., Tantillo, G., Ghibelli, L., Sabbatini, L., Bleve-Zacheo, T., DÁlessio, M., Zambonin, P. G., and Traversa, E. (2005) Copper nanoparticle/polymer composites with antifungal and bacteriostatic properties. *Chemistry of Materials*, **17**, 5255-5262.
84. Gu, H.W., Ho, P.L., Tong, E., Wang, L., and Xu, B. (2003) Presenting vancomycin on nanoparticles to enhance antimicrobial activities. *Nano Letters*, **3**, 1261-1263.
85. Kumar, R., and Munstedt, H. (2005) Silver ion release from antimi-crobial polyamide/silver composites. *Biomaterials*, **26**, 2081-2088.
86. Damm, C., Munsted, H., and Rosch, A. (2008) The antimicrobial effi-cacy of polyamide 6/silver-nano- and microcomposites. *Materials Chemistry and Physics*, **108**, 61-66.
87. Su, H. L., Lin, S. H., Wei, J. C., Pao, I. C., Chiao, S. H., Huang, C. C., et al. (2011) Novel nanohybrids of silver particles on clay platelets for in-hibiting silver-resistant bacteria. *PLoS One*, **6**(6), e21125.
88. Dallas, P., Sharma, V. K., and Zboril, R. (2011) Silver polymeric nano-composites as advanced antimicrobial agents: classification, syn-thetic paths, applications, and perspectives. *Advances in Colloid and Interface Science*, **166**(1-2), 119-135.
89. Sanpui, P., Murugadoss, A., Prasad, P. V. D., Ghosh, S. S., and Chatto-padhyay, A. (2008) The antibacterial properties of a novel chitosan-Ag-nanoparticle composite. *International Journal of Food Microbiol-ogy*, **124**, 142-146.
90. Tankhiwale, R., and Bajpai, S. K. (2009) Graft copolymerization onto cellulose-based filter paper and its further development as silver nanoparticles loaded antibacterial food packaging material. *Colloids and Surfaces*, **69**, 164-168.
91. Tankhiwale, R., and Bajpai, S. K. (2010) Silver-nanoparticle-loaded chitosan lactate films with fair antibacterial properties. *Journal of Applied Polymer Science*, **115**(3), 1894-1900.
92. De Moura, M. R., Mattoso, L. H. C., and Zucolotto, V. (2012) Develop-ment of cellulose-based bactericidal nanocomposites containing sil-ver nanoparticles and their use as active food packaging. *Journal of Food Engineering*, **109**, 520-524.

93. Jin, T., Sun, D., Su, J. Y., Zhang, H., and Sue, H. (2009) Antimicrobial efficacy of zinc oxide quantum dots against *L. monocytogenes, S. enteritidis,* and *E. coli* O157:H7. *Journal of Food Science,* 1, M46-M52.
94. Al-Hazmi, F., Alnowaiser, F., Al-Ghamdi, A. A., Al-Ghamdi, A. A., Aly, M. M., Al-Tuwirqi, R. M., et al. (2012) A new large-scale synthesis of magnesium oxide nanowires: structural and antibacterial properties. *Superlattices and Microstructures,* 52(2), 200-209.
95. Busolo, M. A., and Lagaron, J. M. (2012) Oxygen scavenging polyolefin nanocomposite films containing an iron modified kaolinite ofinterest in active food packaging applications. *Innovative Food Science and Emerging Technologies,* 16, 211-217.
96. Li, X.-e., Green, A. N. M., Haque, S. A., Mills, A., and Durrant, J. R. (2004) Light-driven oxygen scavenging by titania/polymer nanocomposite films. *Journal of Photochemistry and Photobiology A: Chemistry,* 162, 253-259.
97. Maneerat, C., and Hayata, Y. (2008) Gas-phase photocatalytic oxidation of ethylene with TiO2-coated packaging film for horticultural products. *Transactions of the ASABE,* 51, 163-168.
98. Ye, S.-y., Tian, Q.-m., Song, X.-l., and Luo, S.-c. (2009) Photoelectrocatalytic degradation of ethylene by a combination of TiO2 and activated carbon felts. *Journal of Photochemistry and Photobiology A: Chemistry,* 208, 27-35.
99. Yam, K. L., Takhistov, P. T., and Miltz, J. (2005) Intelligent packaging: concepts and applications. *Journal of Food Science,* 70, R1-R10.
100. Kerry, J. P., O'Grady, M. N., and Hogan, S. A. (2006) Past, current and potential utilization of active and intelligent packaging systems for meat and muscle-based products: a review. *Meat Science,* 74, 113-130.
101. O'Grady, M. N., and Kerry, J. P. (2008) Smart packaging technology. In: *Meat Biotechnology,* Toldra, F. (ed.), Springer, USA, p. 425-451.
102. Hogan, S. A., and Kerry, J. (2008) Smart packaging of meat and poultry products. In: *Smart Packaging Technologies for Fast Moving Consumer Goods,* Kerry, J., and Butler, P. (eds.), John Wiley and Sons Ltd., USA.
103. Taoukis, P. S., and Labuza, T. P. (2003) Time-temperature indicators (TTIs). In: *Novel Food Packaging Techniques,* Ahvenainen, R. (ed.), Woodhead Publishing, UK, pp. 103-126.
104. Smolander, M., Hurme, E., Latva-Kala, K., Luoma, T., Alakomi, H. L., and Ahvenainen, R. (2002) Myoglobin-based indicators for the evaluation of freshness of unmarinated broiler cuts. *Innovative Food Science and Emerging Technologies,* 3, 279-288.
105. De Abreu, D. A. P., Cruz, J. M., and Losada, P. P. (2011) Active and intelligent packaging for the food industry. *Food Reviews International,* 28, 146-187.
106. McFarlane, D., and Sheffi, Y. (2003) The impact of automatic identi-

fication on supply chain operations. *International Journal of Logistics Management*, **14**, 1-17.

107. Manthou, V., and Vlachopoulou, M. (2001) Bar-code technology for inventory and marketing management systems: a model for its development and implementation. *International Journal of Production Economics*, **71**, 157-164.

108. Han, J. H. (2013) A review of food packaging technologies and innovations. In: *Innovations in Food Packaging*, Han, J. H. (ed.), Elsevier Academic Press, The Netherlands.

109. Drobnik, O. (2015) *Barcodes with IOS: Bringing Together the Digital and Physical Worlds*, Manning, USA.

110. Yam, K. L., Takhistov, P. T. W., and Miltz, J. W. (2009) Intelligent packaging. In: *The Wiley Encyclopedia of Packaging Technology*, Yam, K. (ed.), 3rd edition, John Wiley and Sons, USA.

111. Kato, H., Tan, K. T., and Chai, D. (2010) *Barcodes for Mobile Devices*, Cambridge University Press, UK.

112. Sarac, A., Absi, N., and Dauzere-Peres, S. (2010) A literature review on the impact of RFID technologies on supply chain management. *International Journal of Production Economics*, **128**, 77-95.

113. Badia-Melis, R., Ruiz-Garcia, L., Garcia-Hierro, J., and Villalba, J. I. R. (2015) Refrigerated fruit storage monitoring combining two different wireless sensing technologies: RFID and WSN. *Sensors*, **15**, 4781-4795.

114. Ariff, M. H., Ismarani, I., and Shamsuddin, N. (2014) RFID Based Systematic Livestock Health Management System. *2014 IEEE Conference on Systems, Process and Control (ICSPC 2014)*, Malaysia, doi: 10.1109/SPC.2014.7086240.

115. Uysal, I., Emond, J., and Bennett, G. (2011) Tag Testing Methodology for RFID Enabled Temperature Tracking and Shelf Life Estimation. *2011 IEEE International Conference on RFID-Technologies and Applications*, Spain, doi: 10.1109/RFID-TA.2011.6068608.

116. Hwang, Y. M., Moon, J., and Yoo, S. (2015) Developing A RFID-based food traceability system in Korea Ginseng Industry: focused on the business process reengineering. *International Journal of Control and Automation*, **8**, 397-406.

117. Bagchi, A. (2012) Intelligent sensing and packaging of foods for enhancement of shelf life: concepts and applications. *International Journal of Scientific and Engineering Research*, **3**(10).

118. Patel, P. D., and Beveridge, C. (2003) In-line sensors for food process monitoring and control. In: *Rapid and On-line Instrumentation for Food Quality Assurance*, Tothill, E. (ed.), Woodhead Publishing, UK.

119. Kress-Rogers, E. (1998) Terms in instrumentation and sensors technology. In: *Instrumentation and Sensors for the Food Industry*, Kress-Rodgers, E. (ed.), Woodhead Publishing, UK, pp. 673-691.

120. Neethirajan, S., Jayas, D. S., and Sadistap, S. (2009) Carbon dioxide

(CO_2) sensors for the agri-food industryeA review. *Food and Bioprocess Technology*, **2**, 115-121.

121. Gao, F., Zheng, D., Tanaka, H., Zhan, F., Yuan, X., Gao, F., and Wang, Q. (2015) An electrochemical sensor for gallic acid based on Fe2O3/electro-reduced graphene oxide composite: estimation for the antioxidant capacity index of wines. *Materials Science and Engineering: C*, **57**, 279-287.

122. Fan, S. H., Shen, J.,Wu, H.,Wang, K. Z., and Zhang, A. G. (2015) A highly selective turnon colorimetric and luminescence sensor based on a triphenylamine-appended ruthenium (II) dye for detecting mercury ion. *Chinese Chemical Letters*, **26**, 580-584.

123. Ibanez, G. A., and Escandar, G. M. (2011) Luminescence sensors applied to water analysis of organic pollutantsdAn update. *Sensors*, **11**, 11081-11102.

124. Ramachandran, G. (2016) *Assessing Nanoparticle Risks to Human Health*, William Andrew, USA.

125. Marrani, D., 2013. Nanotechnologies and Novel Foods in European Law. *Nanoethics* **7**, 177- 188.

3

Nanosensors for Detection of Toxins and Pathogens in Food

B. Ramya Sree and K. Divakar*

Department of Biotechnology, National Institute of Technology Warangal, Warangal, India

Corresponding author: divakar@nitw.ac.in

3.1 Introduction

Food safety and food quality assessment are the major concerns for food supply chains and human health due to rapid change in food recipes and food habits in the globalized food market. In food industry, right from production, processing, distribution and marketing, the food safety must be ensured [1-3]. Food-borne illnesses related with microbial pathogens, their toxins and other contaminants become serious threat to human and animal health. Food products which are in contact with contaminated water sources or fecal matter during the processing stages as well as undercooked products have higher risk of contamination and presence of microbial load. For the last two decades, significant research efforts have been made for the detection of food pathogens as well as toxic metabolites produced by them. Food products of animal origin including meat, poultry and dairy products are the major reservoirs for many food-borne pathogens, including *Salmonella, Campylobacter, Listeria, Escherichia coli, Vibrio* and prions that cause bovine spongiform encephalopathy. Many of these microbes spread infection even at low dosage level, with higher risk for elderly, infants and people with immunological deficiencies [2,4-6].

Significant investments have been made by the food industries to prevent the food-borne pathogens as well as to develop sensitive methods for the detection of pathogens in the food products [7]. Several conventional detection methods exist for the detection of food-borne pathogens and toxins, however, these are time consuming and

Recent Trends in Nanobiotechnology, edited by Prakash Saudagar and K. Divakar
© 2019 Central West Publishing, Australia

laborious. Thus, there is a need of sophisticated instruments and trained personnel for the detection of pathogens at critically low concentrations [6,8-10]. In recent years, nanotechnology has been an emerging platform for solving the issue of food-borne pathogen detection [11-14]. Thus, a suitable alternative for the existing traditional pathogen detection methods is the development of sensors based on nanomaterials (physical and chemical sensors), which can be used off-line or online and integrated with the existing food processing equipment for food contaminant detection.

3.2 Magnetic Nanoparticles (MNPs) as Bacterial Sensors

Use of MNPs for detecting and capturing bacteria has certain essential dimensions like size similarity to bacteria, magnetic conduct and biomolecular conjugation possibilities. Several studies have exploited MNPs for effective capture of pathogens [15-32].

Lin *et al.* [18] developed vancomycin-modified MNPs to trap Gram-positive bacteria selectively. Later, isolated cells were characterized using MALDI-MS, thus, effectively reducing the interference of proteins and other metabolites in the mass spectra. Yitzhaki *et al.* [17] employed silica magnetic particles coated with didecyldimethylammonium bromide for concentrating and capturing spores of *Bacillus* to achieve rapid, simple and sensitive detection. Chen *et al.* [19] devised titania-coated iron oxide MNPs that bind to bacteria under magnetic field, thus, concentrating them for further identification and analysis. Gram-negative bacteria like *E. coli* O157:H7, uropathogenic *E. coli*, *Pseudomonas aeruginosa*, *Klebsiella pneumoniae* and *Shigella sonnei* were swiftly identified from the samples using this approach.

Kell *et al.* [20] designed superparamagnetic NPs conjugated with polystyrene beads and studied the factors affecting the specificity and efficiency of bacterial capture. Ryan *et al.* [23] employed a novel superparamagnetic NP conjugate having a single-domain antibody specific for *S. aureus* with high sensitivity and specificity. Huang *et al.* [33] applied amine-functionalized MNPs, which could non-specifically capture and remove a wide range of pathogenic bacteria from different water, grape juice and green tea samples.

Kim *et al.* [12] designed an enhanced colorimetric method that employed immunomagnetic separation for the concentration of *E. coli* O157:H7 with the help of anti-strain antibodies and streptavidin-alkaline phosphatase coated onto magnetic NPs. Enzyme-catalyzed precipitation reaction resulted in an amplified signal for the detection. This technique was applied to detect bacteria in an artificially fortified lettuce sample, which showed no cross-reactivity towards non-target species, proving its potential.

3.3 Nano-biosensors

3.3.1 Immuno-based Nanosensors

The standard enzyme-linked immuno-sorbent assay (ELISA) uses antigen-antibody interactions to detect the presence or absence of a counter molecule. Bacteria are also detected using a similar technique by Swanink *et al.* [34]. A plethora of sensing systems have been devised to integrate antibody based platforms to achieve portability, reduced analysis time and ease of detection [35]. Gold nanoparticles (AuNP) based nano-sensing systems transformed with antibodies have been largely formulated, and detection systems based on surface plasmon resonance (SPR) measurement or color change due to aggregation/disaggregation when bound to the target are incorporated. Baccar *et al.* [36] designed an immunosensor to detect *E. coli* K12 and *Lactobacillus fermentium* using SPR that had immobilized antibodies specific for the bacteria, over a gold layer or AuNPs deposited on the gold layer. Changes in the resonance angle and refractive index were measured with different bacterial concentrations. An immunosensor having *E. coli* antibody coated gold nanorods and two-photon Rayleigh scattering (TPRS) spectroscopy for detection of *E. coli* has been reported by Singh *et al.* [37]. The functionalized nanorods could bind to *E. coli* O157:H7 bacteria, leading to their aggregation, resulting in enhanced signal for detection. *Giardia lamblia* cysts were colorimetrically detected using antibody coated AuNP probes that change color due to binding, showing increased absorbance at 550 nm when the concentration of bacteria increased [38].

An electrochemical immunosensor having trifunctional nanoparticles with immuno-magnetic/polyaniline core/shell to achieve specificity, improved separation and enhanced conductivity for detection (which detects *Bacillus spp.* and *E. coli* O157:H7 using cyclic voltammetry and amperometry as the detection systems) has also been designed [39]. Cost-effective paper-based systems have also been exploited where nitrocellulose paper modified with antibodies against bacteria and AuNPs were used for detection. A portable multiplex strip reader that had bacterial antibodies bound to AuNPs coated on the nitrocellulose paper was devised for the detection of *Staphylo-coccus aureus* and *Pseudomonas aeruginosa* [40]. Inbaraj *et al.* [41] and Ye *et al.* [42] designed electrochemical immunosensors having graphene quantum dots to detect *Salmonella*. Presence of bacteria in the food and environmental samples was also shown to be detected by Raman spectroscopy of nano-modified surfaces [43,44]. Mathelié-Guinlet *et al.* [45] designed an *E. coli* detection system that employed immobilized silica NPs. Molecular recognition and selectivity were greatly enhanced due to the use of immobilized sensors on various nanomaterials and 3D modelling [46-53]. Overall, immunosensors have been developed to improve the efficiency of pathogen detection, thus, resolving a large number of bioassay challenges [54]. However, critical issues related to antibody orientation and adsorption speci-ficity still need attention.

3.3.2 Nanosensors based on Bacteriophages

Bacteriophages are viruses that can identify a bacteria, infect it and lyse it after amplifying themselves [55-60]. They have an inherent ability to recognize and differentiate strains of bacteria [4], even live or dead cells [5], and can be produced in large quantities at reasona-ble costs [55,56], thus, making them good candidates as recognition molecules in biosensor design for detecting bacteria (Table 3.1). A multifunctional carboxyl functionalized magnetic probe conjugated to amino-modified T7 phage was employed in the detection, concen-tration and separation of *E. coli* from drinking water [57]. Cells were lysed after identification and magnetic separation, releasing β-galac-tosidase, which was assayed by UV spectrometry and visualized by a

Table 3.1 Phage based nanosensors

Bacteria	Bacterio-phage employed	Sample source	Detection system	Ref.
E. coli K12	M13KE	Water, or-ange juice and skimmed milk	Colorimetry	[58]
E. coli K12	T4	Milk	Electrochemical/ impedimetric	[60]
E. coli	T7	Drinking water	Optical/colorim-etry	[57]

mobile camera for analysis. Other studies also described designs that use optical transduction for analysis [57-69].

Adsorbed phages have been used in a biosensor that detects *Salmonella typhimurium* using a piezoelectric transducer monitoring the decrease in the resonance frequency [69]. Tawil *et al.* [5] developed a T4 phage sensor to detect *E. coli* O157: H7 and a novel B14 phage to detect methicillin resistant *Staphylococcus aureus* [5]. Non-specific adsorption was prevented by adding BSA, and lysis of bacteria on contact with phage caused quantitative change in the SPR signals. Another study by Guntupalli *et al.* [70] employed immobilized lytic phages to detect and differentiate methicillin sensitive and resistant *Staphylococcus aureus* using a quartz crystal microbalance dissipation, where phage bound bacteria had more dissipation energy and less frequency. Strains were easily distinguished as these interacted differently to the antibody of penicillin-binding protein. Neufeld *et al.* [71] designed a phage-sensing system that used electrochemical quantification of β-D-Galactosidase enzyme in the lysate of *E.coli* on reaction with the substrate p-aminophenyl-β-D-galactopyranoside, forming p-aminophenol with the amperometric sensors. Phages are also used as recognition probes that measure direct impedance of bacteria in a sample [5,60,69,70]. Shabani *et al.* [60] used magnetic bead-conjugated T4 phage immobilized on carbon electrode microarrays as a probe and quantified bacteria with impedance spectroscopy. Table 3.1 presents other nanosensors for microbial detection.

3.3.3 Nanosensors based on Aptamers

Aptamers are defined as the short oligos of nucleotides or peptides synthesized by in-vitro evolution of ligands [72-74]. They have a wide scope of use as biosensors in analytical assays due to high affinity and specificity to their targets [72,75,76]. Aptamer based biosensors are designed using single-walled [66]/multi-walled [61] carbon nanotubes (CNTs), graphene oxide, fluorescent quantum dots [58,77], AuNPs [78] and magnetic beads [79] for a variety of purposes. Single-walled CNTs functionalized with aptamers have been designed for the potentiometric detection of *E.coli* [75]. Their assembly was studied with molecular dynamics simulations by Abbaspour *et al.* [66]. Spontaneous change in conformation of DNA was observed by Johnson *et al.* [80] that enables self-assembly of CNTs. Change in conformation of the functionalized aptamer, in the presence of bacteria, induces a change in the surface charge and potential, leading to their detection [81]. Zelada-Guillén *et al.* [61] designed a potentiometric CNT aptasensor that selectively detected and differentiated *E. coli*'s non-pathogenic strain CECT 675 and pathogenic O157:H7 from milk and apple juice [61]. Table 3.2 also demonstrates some examples of aptamer based nanosensors.

Table 3.2 Aptamer based nanosensors

Bacteria	NPs employed	Sample source	Detection system	Ref.
Non- pathogenic *E. coli* CECT 675 and pathogenic *E.coli* O157:H7	Single walled CNTs	Milk and apple juice	Potentiometry	[61]
Salmonella typhimurium and *Vibrio parahaemolyticus*	Carbon dots	Shrimps	Optical/fluorescence	[62]
Staphylococcus aureus	Nanocomposite of graphene oxide and AuNPs	Water and fish	Electrochemical/ impedance	[14]
Staphylococcus aureus, Salmonella typhimurium and *Vibrio parahaemolyticus*	Rare-earth UCNPs and Fe_2O_3 MNPs	Milk and shrimp	Optical/luminescence	[63]
Salmonella enterica serovar Typhimurium	AuNPs conjugated to antibodies and horse radish peroxidase	Milk	Optical	[11]
Salmonella typhimurium and *Vibrio parahaemolyticus*	QDs and novel amorphous carbon NPs	Chicken and shrimp	Optical/dual-FRET	[64]
Staphylococcus aureus	Graphene interdigitated Au electrode	Milk	Mechanical/piezoelectric quartz crystal electrode	[65]

Staphylococcus aureus	AgNPs	Water	Electrochemical/ stripping voltammetry	[66]
Salmonella sp.	Multi-walled CNTs	chicken	Amperometric-cyclic voltammetry and impedimetric	[67]
E.coli, Bacillus subtilis, Bacillus atrophaeus and *Listeria innocua*	Enzyme-DNA conjugate array chip	Meat juice	Electrochemical	[68]

Electrochemical sensors based on carbon nanomaterials have also been developed for pathogen detection. Zuo *et al.* [82] integrated graphene oxide coated aptasensors with paper/glass/polydimethylsiloxane for simultaneous rapid multiplex detection of *Lactobacillus acidophilus, Staphylococcus aureus* and *Salmonella enterica* by fluorescence quenching. An impedimetric apta-biosensor having graphene oxide and AuNPs was fabricated by Jia *et al.* [14] for the detection of *S. aureus*. Yoo *et al.* [83] achieved the detection of *Salmonella typhimurium, Lactobacillus acidophilus* and *Pseudomonas aeruginosa* by thiol immobilized specific aptamers on a multispot array chip coated with AuNPs, measuring the localized SPR changes when bound by bacteria. A piezoelectric gold aptasensor was designed by Lian *et al.* [65] to detect *S. aureus* using mercaptobenzenediazonium tetrafluoroborate (MBDT) functionalized graphene. When the aptamer binds to the bacteria, it detaches from the graphene surface leading to variation in the oscillator frequency. Jia *et al.* [67] fabricated *Salmonella* specific amino-modified aptasensors functionalized on reduced graphene oxide and carboxyl-altered multi-walled CNTs, when exposed to bacteria, led to difference in impedance due to blockage of electron transfer.

Colorimetric detection of *E. coli* O157:H7 and *Salmonella typhimurium* using AuNP functionalized aptamers was developed by Wu *et al.* [84]. NPs aggregate in the presence of specific bacteria resulting in the shift of wavelength from red to blue, which can be quantified. This strategy can lead to the development of simple and portable systems that can quantify results with a color quantifying software. An abundant diversity of gold nanostructures like nanorods [85], nanostars [13], nanocrystals [38] and aptamer-altered AuNPs [11,86,87] has been employed in bacterial estimation. Kang *et al.* [88] fabricated an electrochemical immuno-sensor that detected *Bacillus cereus* using a glassy carbon electrode coated with a double-layer of chitosan and AuNPs. Use of nanoscale polydiacetylene vesicles surface coated with lipopolysaccharide-binding aptamers have also been reported by Wu *et al.* [89].

Fluorescent markers such as cadmium-telluride quantum dots (CdTe QDs) have been coupled with aptamers for selectivity, recognition and detection [62,64]. Tawil *et al.* [5] devised aptamer-modified quantum dots that specifically recognize *Salmonella typhimurium* and *Vibrio parahaemolyticus* from shrimp samples using flow cytometry. Duan *et al.* [64] detected these two bacteria simultaneously with green and red-emitting quantum dots along with amorphous carbon NPs as counterparts using fluorescence resonance energy transfer (FRET). Carbon NPs quench the fluorescence of quantum dots when the target bacteria are absent, and the quenching is suppressed to emit light based on their concentration in the presence of bacteria.

Up conversion NPs (UCNPs) having rare earth metals have also been used as luminescent aptamer labels for multiplexed analysis of *Vibrio parahaemolyticus*, *Salmonella typhimurium* and *Staphylococcus aureus* by Wu *et al.* [63]. For a specific bacteria, aptamer is coupled with particular UCNPs, magnetic NPs are coupled with specific cDNA, and, in turn, aptamer-UCNPs are conjugated to cDNA-magnetic NPs. In the presence of bacteria, the aptamer dissociates from UCNPs and attaches to bacteria, thus, quenching the emission peak. Li *et al.* [90] and Liao *et al.* [91] reported the use of different fluorescent labels for the detection of pathogens.

Immunomagnetic platforms have been employed in conjunction with aptamer nanosensors for the detection and separation of bacteria using selectivity and electrochemical detection systems respectively. In a recent study, Abbaspour *et al.* [66] employed a dual aptamer sandwich system, where one aptamer was conjugated to AgNPs and other was conjugated to magnetic beads, for the detection of *S. aureus*. One of the aptamers was anti-*S. aureus* aptamer and the other was used for the quantification of the signal using differential pulse stripping voltammetry. Also, *Salmonella typhimurium* and *Vibrio parahaemolyticus* were simultaneously detected using carbon dots and florescent aptasensors [64,92]. In another study, Liu *et al.* [93] devised an electrochemical aptasensor having nanostructured gold microelectrode manufactured by the deposition of dendritic gold structures.

3.3.4 Nanosensors based on Peptides (Anti-microbial Peptides)

Anti-microbial peptides (AMPs) are oligos that act as a host's defense against invading species [94]. They bind to the surface of the invaders, thus, disrupting their cell membrane [95]. AMPs are widely used in clinical sector for the detection of infections [96,97]. Many research studies have also analyzed the properties of natural and synthetic AMPs as recognition molecules for the detection and differentiation of bacteria [98-101]. Bacterial binding to the surface of the electrodes is usually monitored with impedance spectroscopy. The pros of using these molecules are better activity and stability even when the environment is harsh, but lack of selectivity is their limitation. However, synthetic AMPs can be designed rationally to possess better binding characteristics, low production cost and high stability [102-105]. Liu *et al.* [106] designed an impedimetric sensor for the detection of *E. coli*, *Pseudomonas aeruginosa*, *Staphylococcus aureus* and *Staphylococcus epidermidis* with the engineered supramolecular AMPs conjugated to AuNP- functionalized electrode in a specific orientation [106]. Peptides are cysteine-modified to bind to the gold surface due to affinity. Rapid label-free detection is possible, enabling high-throughput screening using chip based sensors in the near future [107]. This approach can also be extended to detect and prevent biofilm formation by modifying the surfaces.

Magnetic NPs coated with proteases have also been explored by different research communities. *Listeria monocytogenes* has been detected using its protease-specific substrate coated onto magnetic NPs that cleaves its proteases, which is further detected colorimetrically by the change of color from black to golden upon action in the artificially fortified ground meat and milk [108]. Other studies have also reported paper-based sensors that detect enzyme activity. Change in color of the substrate due to the enzyme release was estimated in these cases [109,110]. Adkins *et al.* [110] designed a colorimetric and electrochemical detection system which used differential enzyme production by *E. coli* and *Enterococcus spp.* as a tool to identify the bacteria, based on the enzymatic conversion of p-nitrophenyl--D-glucuronide into p-nitrophenol at pH above 7.2.

3.4 3D Printing Technologies for Nanobiosensors

Biosensing can reach newer heights with the latest technology of 3D printing of sensor prototypes and organization of sensing layers

[111,112]. With the availability of new printing technologies, materials and processes, new avenues of biotechnological applications like 3D cell cultures, tissue engineering [113,114] and biosensor designs [115] have blossomed. Few research studies have proved these possibilities [116,117].

Klein *et al.* [118] printed 3D objects with the ink made of biomolecule conjugated photopolymers. Diagnostic devices like immunosensors and enzymatic 3D objects used for biosensing and catalysis are some of the success stories in 3D printing. Mandon *et al.* [112] formulated 4D printed objects with hydrogel sensing layers which could recognize and catalyze as well as have fluidic abilities for biosensing and biochip applications. For this purpose, a printing technology called Digital Light Processing was used, which could synthesize objects of good resolution. These objects had advantages like improved biorecognition capabilities and possession of multifunctional complexes that allow flexibility and wider scope of use.

3.5 Conclusions

Methods used for the detection of microbial pathogens and toxins vary based on the capital cost, size of detector, response time, sensitivity and selectivity. Smart nanosensors are reported to quantify pathogenic microbes at lower microbial count and toxins at molecular levels. These sensors offer possibilities of integration with the existing process instruments and online detection in the food processing industries. Overall, ease of handling and monitoring make nanotechnology based detection systems promising candidates for the development of pathogen and toxin detection methods in food processing industries in the near future.

References

1. Wahidin, D., and Purnhagen, K. (2018) Improving the level of food safety and market access in developing countries. *Heliyon*, **4**(7), e00683.
2. Fung, F., Wang, H.-S., and Menon, S. (2018) Food safety in the 21st century. *Biomedical Journal*, **41**(2), 88-95.
3. Batt, C. A. (2016) Food Safety Assurance. *Reference Module in Food Science*, doi: 10.1016/B978-0-08-100596-5.03442-9.
4. Baggesen, D. L., and Wegener, H. C. (1994) Phage types of Salmonella enterica ssp. enterica serovar typhimurium isolated from pro-

duction animals and humans in Denmark. *Acta Veterinaria Scandinaica*, **35**(4), 349-354.

5. Tawil, N., Sacher, E., Mandeville, R., and Meunier, M. (2012) Surface plasmon resonance detection of E. coli and methicillin-resistant S. aureus using bacteriophages. *Biosensors & Bioelectronics*, **37**(1), 24-29.

6. Mandal, P. K., Biswas, A. K., Choi, K., and Pal, U. K. (2011) Methods for rapid detection of foodborne pathogens: An overview. *American Journal of. Food Technology*, **6**(2), 87-102.

7. Safavieh, M., Nahar, S., Zourob, M., and Ahmed, M. U. (2015) Microfluidic biosensors for high throughput screening of pathogens in food. In: *High Throughput Screening for Food Safety Assessment*, Bhunia, A. K., Kim, M. S., and Taitt, C. R. (eds.), pp. 327-357.

8. Kaspar, C. W., and Tartera, C. (1990) Methods for detecting microbial pathogens in food and water. *Methods in Microbiology*, **22**, 497-531.

9. Pinu, F. R. (2016) Early detection of food pathogens and food spoilage microorganisms: Application of metabolomics. *Trends in Food Science and Technology*, **54**, 213-215.

10. Böhme K., Antelo, S. C., Fernandez-No, I. C., Quintela-Baluja, M., Barros-Velazquez, J., Canas, B., and Calo-Mata, P. (2016) Detection of foodborne pathogens using MALDI-TOF mass spectrometry. In: *Antimicrobial Food Packaging*, Barros-Velazquez, J. (ed.), Academic Press, USA, pp. 203-214.

11. Wu W., Li, J., Pan, D., Li, J., Song, S., Rong, M., Li, Z., Gao, J., and Lu, J. (2014) Gold nanoparticle-based enzyme-linked antibody-aptamer sandwich assay for detection of salmonella typhimurium. *ACS Applied Materials & Interfaces*, **6**(19), 16974-16981.

12. Kim, S. U., Jo, E. J., Mun, H., Noh, Y., and Kim, M. G. (2018) Ultrasensitive detection of Escherichia coli O157:H7 by immunomagnetic separation and selective filtration with nitroblue tetrazolium/5-bromo-4-chloro-3-indolyl phosphate signal amplification. *Journal of Agricultural and Food Chemistry*, **66**(9), 4941-4947.

13. Verma, M. S., Chen, P. Z., Jones, L., and Gu, F. X. (2014) Branching and size of CTAB-coated gold nanostars control the colorimetric detection of bacteria. *RSC Advances*, **4**, 10660-10668.

14. Jia F., Duan, N., Wu. S., Ma, X., Xia, Y., Wang, Z., and Wei, X. (2014) Impedimetric aptasensor for Staphylococcus aureus based on nanocomposite prepared from reduced graphene oxide and gold nanoparticles. *Microchimica Acta*, **181**, 967-974.

15. Jinhao, G. A. O., Hongwei, G. U., and Bing, X. U. (2009) Multifunctional magnetic nanoparticles: Design, synthesis, and biomedical applications. *Accounts of Chemical Research*, **42**, 1097-1107.

16. Shan Z., Wu, Q., Wang, X., Zhou, Z., Oakes, K. D., Zhang, X., Huang, Q., and Yang, W. (2010) Bacteria capture, lysate clearance, and plasmid

DNA extraction using pH-sensitive multifunctional magnetic nanoparticles. *Analytical Biochemistry*, **398**, 120-122.

17. Yitzhaki, S., Zahavy, E., Oron, C., Fisher, M., and Keysary, A. (2006)
Concentration of Bacillus spores by using silica magnetic particles.
Analytical Chemistry, **78**, 6670-6673.

18. Lin, Y. S., Tsai, P. J., Weng, M. F., and Chen, Y. C. (2005) Affinity capture using vancomycin-bound magnetic nanoparticles for the
MALDI-MS analysis of bacteria. *Analytical Chemistry*, **77**, 1753-
1760.

19. Chen, W. J., Tsai, P. J., and Chen, Y. C. (2008) Functional nanoparticle-
based proteomic strategies for characterization of pathogenic bacteria. *Analytical Chemistry*, **80**, 9612-9621.

20. Kell, A. J., Somaskandan, K., Stewart, G., Bergeron, M. G., and Simard,
B. (2008) Superparamagnetic nanoparticle-polystyrene bead conjugates as pathogen capture mimics: A parametric study of factors affecting capture efficiency and specificity. *Langmuir*, **24**, 3493-3502.

21. Bromberg, L., Raduyk, S., and Hatton, T. A. (2009) Functional magnetic nanoparticles for biodefense and biological threat monitoring
and surveillance. *Analytical Chemistry*, **81**, 5637-5645.

22. Zhao, X., Hilliard, L. R., Mechery, S. J., Wang, Y., Bagwe, R. P., Jin, S.,
and Tan, W. (2004) A rapid bioassay for single bacterial cell quanti-
tation using bioconjugated nanoparticles. *Proceedings of the National Academy of Sciences of the United States of America*, **101**,
15027-15032.

23. Ryan S., Kell, A. J., van Faassen, H., Tay, L.-L., Simard, B., MacKenzie,
R., Gilbert, M., and Tanha, J. (2009) Single-domain antibody-nano-
particles: Promising architectures for increased Staphylococcus aureus detection specificity and sensitivity. *Bioconjugate Chemistry*,
20, 1966-1974.

24. Ho, K. C., Tsai, P. J., Lin, Y. S., and Chen, Y. C. (2004) Using biofunc-
tionalized nanoparticles to probe pathogenic bacteria. *Analytical
Chemistry*, **76**, 7162-7168.

25. Liu, J. C., Tsai, P. J., Lee, Y. C., and Chen, Y. C. (2008) Affinity capture
of uropathogenic Escherichia coli using pigeon ovalbumin-bound
$Fe_3O_4@Al_2O_3$ magnetic nanoparticles. *Analytical Chemistry*, **80**,
5425-5432.

26. El-Boubbou, K., Gruden, C., and Huang, X. (2007) Magnetic glyco-nanoparticles: A unique tool for rapid pathogen detection, decontam-
ination, and strain differentiation. *Journal of American Chemical Society*, **129**, 13392-13393.

27. Lin C. C., *et al.*, (2002) Selective binding of mannose-encapsulated
gold nanoparticles to type 1 pili in Escherichia coli. *Journal of American Chemical Society*, **124**, 3508-3509.

28. Liu, L. H., Dietsch, H., Schurtenberger, P., and Yan, M. (2009) Pho-
toinitiated coupling of unmodified monosaccharides to iron oxide

nanoparticles for sensing proteins and bacteria. *Bioconjugate Chemistry*, **20**, 1349-1355.

29. Gu, H., Ho, P. L., Tsang, K. W. T., Wang, L., and Xu, B. (2003) Using Biofunctional Magnetic nanoparticles to capture vancomycin-resistant enterococci and other gram-positive bacteria at ultralow concentration. *Journal of American Chemical Society*, **125**, 15702-15703.

30. Wu, L., Mendoza-Garcia, A., Li, Q., and Sun, S. (2016) Organic phase syntheses of magnetic nanoparticles and their applications. *Chemical Reviews*, **116**, 10473-10512.

31. Kell A. J., Stewart, J., Ryan, S., Peytavi, R., Boissinot, M., Huletsky, A., Bergeron, M. G., and Simard, B. (2008) Vancomycin-modified nanoparticles for efficient targeting and preconcentration of gram-positive and gram-negative bacteria. *ACS Nano*, **2**, 1777-1788.

32. Honda, H., Kawabe, A., Shinkai, M., and Kobayashi, T. (1998) Development of chitosan-conjugated magnetite for magnetic cell separation. *Journal of Fermentation and Bioengineering*, **86**, 191-196.

33. Huang, Y. F., Wang, Y. F., and Yan, X. P. (2010) Amine-functionalized magnetic nanoparticles for rapid capture and removal of bacterial pathogens. *Environmental Science & Technology*, **44**(20), 7908-7913.

34. Swanink, C. M. A., Meis, J. F. G. M., Rijs, A. J. M. M., Donnelly, J. P., and Verweij, P. E. (1997) Specificity of a sandwich enzyme-linked immunosorbent assay for detecting Aspergillus galactomannan. *Journal of Clinical Microbiology*, **35**(1), 257-260.

35. Güner, A., Çevik, E., Şenel, M., and Alpsoy, L. (2017) An electrochemical immunosensor for sensitive detection of Escherichia coli O157:H7 by using chitosan, MWCNT, polypyrrole with gold nanoparticles hybrid sensing platform. *Food Chemistry*, **229**, 358-365.

36. Baccar, H., Mejri, M. B., Hafaiedh, I., Ktari, T., Aouni, M., and Abdelghani, A. (2010) Surface plasmon resonance immunosensor for bacteria detection. *Talanta*, **82**(2), 810-814.

37. Singh A. K., Senapati, D., Wang, S., Griffin, J., Neely, A., Candice, P., Naylor, K. M., Varisli, B., Kalluri, J. R., and Ray, P. C. (2009) Gold nanorod based selective identification of escherichia coli bacteria using two-photon Rayleigh scattering spectroscopy. *ACS Nano*, **3**(7), 1906-1912.

38. Li, X. X., Cao, C., Han, S. J., and Sim, S. J. (2009) Detection of pathogen based on the catalytic growth of gold nanocrystals. *Water Research*, **43**(5), 1425-1431.

39. Setterington, E. B., and Alocilja, E. C. (2012) Electrochemical biosensor for rapid and sensitive detection of magnetically extracted bacterial pathogens. *Biosensors*, **2**(1), 15-31.

40. Li, C.-z., Vandenberg, K., Prabhulkar, S., Zhu, X., Schneper, L., Methee, K., Rosser, C. J., and Almeide, E. (2011) Paper based point-of-care

testing disc for multiplex whole cell bacteria analysis. *Biosensors and Bioelectronics,* **26**(11), 4342-4348.

41. Inbaraj, B. S., and Chen, B. H. (2016) Nanomaterial-based sensors for detection of foodborne bacterial pathogens and toxins as well as pork adulteration in meat products. *Journal of Food and Drug Analysis,* **24**(1), 15-28.

42. Ye W., Guo, J., Bao, X., Chen, T., Weng, W., Chen, S., and Yang, M. (2017) Rapid and sensitive detection of bacteria response to antibiotics using nanoporous membrane and graphene quantum dot (GQDs)-based electrochemical biosensors. *Materials,* **10**(6), doi: 10.3390/ma10060603.

43. Li, Y. S., and Church, J. S. (2014) Raman spectroscopy in the analysis of food and pharmaceutical nanomaterials. *Journal of Food Drug Analysis,* **22**(1), 29-48.

44. Deisingh A. K., and Thompson, M. (2004) Biosensors for the detection of bacteria. *Canadian Journal of Microbiology,* **50**(2), 69-77.

45. Mathelié-Guinlet, M., Gammoudi, I., Beven, L., Moroté, F., Delville, M.-H., Grauby-Heywang, C., and Cohen-Bouhacina, T. (2016) Silica nanoparticles assisted electrochemical biosensor for the detection and degradation of escherichia coli bacteria. *Procedia Engineering,* **168**, 1048-1051.

46. Zhang, Y., and Wei, Q. (2016) The role of nanomaterials in electroanalytical biosensors: A mini review. *Journal of Electroanalytical Chemistry,* **781**, 401-409.

47. Wang, Z., Yu, J., Gui, R., Jin, H., and Xia, Y. (2016) Carbon nanomaterials-based electrochemical aptasensors. *Biosensors and Bioelectronics,* **79**, 136-149.

48. Velu, R., Frost, N., and DeRosa, M. C. (2015) Linkage inversion assembled nano-aptasensors (LIANAs) for turn-on fluorescence detection. *Chemical Communications,* **51**(76), 14346-14349.

49. Rowland, C. E., Brown, C. W., Delehanty, J. B., and Medintz, I. L. (2016) Nanomaterial-based sensors for the detection of biological threat agents. *Materials Today,* **19**(8), 464-477.

50. Wang, H., Yang, R., Yang, L., and Tan, W. (2009) Nucleic acid conjugated nanomaterials for enhanced molecular recognition. *ACS Nano,* **3**(9), 2451-2460.

51. Liu, Y., and Yu, J. (2016) Oriented immobilization of proteins on solid supports for use in biosensors and biochips: a review. *Microchimica Acta,* **183**(1), 1-19.

52. Korayem, M. H., Estaji, M., and Homayooni, A. (2017) Noncalssical multiscale modeling of ssDNA manipulation using a CNT-nanocarrier based on AFM. *Colloids and Surfaces B: Biointerfaces,* **158**, 102-111.

53. Jeddi, I., and Saiz, L. (2017) Three-dimensional modeling of single stranded DNA hairpins for aptamer-based biosensors. *Scientific Re-*

ports, **7**, 1178.
54. Khonsari, Y. N., and Sun, S. (2017) Recent trends in electrochemiluminescence aptasensors and their applications. *Chemical Communications*, **53**(65), 9042-9054.
55. Tawil, N., Sacher, E., Mandeville, R., and Meunier, M. (2014) Bacteriophages: Biosensing tools for multi-drug resistant pathogens. *Analyst*, **139**(6), 1224-1236,
56. Guttman, B., Raya, R., and Kutter, E. (2004) Basic phage biology. In *Bacteriophages: Biology and Applications*, Kutter, E., and Sulakvelidze, A. (eds.), CRC Press, USA, Chapter 3.
57. Chen, J., Alcaine, S. D., Jiang, Z., Rotello, V. M., and Nugen, S. R. (2015) Detection of escherichia coli in drinking water using T7 bacteriophage-conjugated magnetic probe. *Analytical Chemistry*, **87**, 8977-8984.
58. Derda, R., Lockett, M. R., Tang, S. K., Fuller, R. C., Maxwell, E. J., Breiten, B., Cuddemi, C. A., Ozdogan, A., and Whitesides, G. M. (2013) Filter-based assay for Escherichia coli in aqueous samples using bacteriophage-based amplification. *Analytical Chemistry*, **85**, 7213-7220.
59. Vinay, M., Franche, N., Grégori, G., Fantino, J. R., Pouillot, F., and Ansaldi, M. (2015) Phage-based fluorescent biosensor prototypes to specifically detect enteric bacteria such as E. coli and Salmonella enterica Typhimurium. *PLoS One*, **10**(7), e0131466.
60. Shabani, A., Marquette, C. A., Mandeville, R., and Lawrence, M. F. (2013) Magnetically-assisted impedimetric detection of bacteria using phage-modified carbon microarrays. *Talanta*, **116**, 1047-1053.
61. Zelada-Guillén, G. A., Bhosale, S. V., Riu, J., and Rius, F. X. (2010) Real-time potentiometric detection of bacteria in complex samples. *Analytical Chemistry*, **82**, 9254-9260.
62. Duan N., Wu, S., Yu, Y., Ma, X., Xia, Y., Chen, X., Huang, Y., and Wang, Z. (2013) A dual-color flow cytometry protocol for the simultaneous detection of Vibrio parahaemolyticus and Salmonella typhimurium using aptamer conjugated quantum dots as labels. *Analytica Chimica Acta*, **804**, 151-158.
63. Wu, S., Duan, N., Shi, Z., Fang, C., and Wang, Z. (2014) Simultaneous aptasensor for multiplex pathogenic bacteria detection based on multicolor upconversion nanoparticles labels. *Analytical Chemistry*, **86**, 3100-3107.
64. Duan, N., Wu, S., Dai, S., Miao, T., Chen, J., and Wang, Z. (2015) Simultaneous detection of pathogenic bacteria using an aptamer based biosensor and dual fluorescence resonance energy transfer from quantum dots to carbon nanoparticles. *Microchimica Acta*, **182**, 917-923.
65. Lian, Y., He, F., Wang, H., and Tong, F. (2015) A new aptamer/graphene interdigitated gold electrode piezoelectric sensor for rapid

                                           Recent Trends in Nanobiotechnology

and specific detection of Staphylococcus aureus. *Biosensors & Bioelectronics*, **65**, 314-319.

66. Abbaspour, A., Norouz-Sarvestani, F. Noori, A., and Soltani, N. (2015) Aptamer-conjugated silver nanoparticles for electrochemical dual-aptamer-based sandwich detection of staphylococcus aureus. *Biosensors & Bioelectronics*, **68**, 149-155.

67. Jia, F., Duan, N., Wu, S., Dai, R., Wang, Z., and Li, X. (2016) Impedimetric Salmonella aptasensor using a glassy carbon electrode modified with an electrodeposited composite consisting of reduced graphene oxide and carbon nanotubes. *Microchimica Acta*, **183**, 337-344.

68. Pöhlmann, C., Wang, Y., Humenik, M., Heidenreich, B., Gareis, M., and Sprinzl, M. (2009) Rapid, specific and sensitive electrochemical detection of foodborne bacteria. *Biosensors & Bioelectronics*, **24**, 2766-2771.

69. Olsen, E. V., Sorokulova, I. B., Petrenko, V. A., Chen, I. H., Barbaree, J. M., and Vodyanoy, V. J. (2006) Affinity-selected filamentous bacteriophage as a probe for acoustic wave biodetectors of Salmonella typhimurium. *Biosensors & Bioelectronics*, **21**, 1434-1442.

70. Guntupalli, R., Sorokulova, I., Olsen, E., Globa, L., Pustovyy, O., and Vodyanoy, V. (2013) Biosensor for detection of antibiotic resistant staphylococcus bacteria. *Journal of Visualized Experiments*, doi: 10.3791/50474.

71. Neufeld, T., Schwartz-Mittelmann, A., Biran, D., Ron, E. Z., and Rishpon, J. (2003) Combined phage typing and amperometric detection of released enzymatic activity for the specific identification and quantification of bacteria. *Analytical Chemistry*, **75**, 580-585.

72. Jayasena, S. D. (1999) Aptamers: An emerging class of molecules that rival antibodies in diagnostics. *Clinical Chemistry*, **45**, 1628-1650.

73. Tuerk, C., and Gold, L. (1990) Systematic evolution of ligands by exponential enrichment: RNA ligands to bacteriophage T4 DNA polymerase. *Science*, **249**, 505-510.

74. Zayats, M., Huang, Y., Gill, R., Ma, C. A., and Willner, I. (2006) Label-free and reagentless aptamer-based sensors for small molecules. *Journal of American Chemical Society*, **128**, 13666-13667.

75. Zelada-Guillen, G. A., Riu, J., Düzgün, A., and Rius, F. X. (2009) Immediate detection of living bacteria at ultralow concentrations using a carbon nanotube based potentiometrie aptasensor. *Angewandte Chemie International Edition*, **48**, 7334-7337.

76. O'Sullivan, C. K. (2002) Aptasensors - The future of biosensing? *Analytical and Bioanalytical Chemistry*, **372**, 44-48.

77. Liu, J., Cheng, J., and Zhang, Y. (2013) Upconversion nanoparticle based LRET system for sensitive detection of MRSA DNA sequence. *Biosensors & Bioelectronics*, **43**, 252-256.

78. Franche, N., Vinay, M., and Ansaldi, M. (2017) Substrate-independ-

ent luminescent phage-based biosensor to specifically detect enteric bacteria such as E. coli. *Environmental Science and Pollution Research International*, **24**, 42-51.

79. De Azeredo, H. M. C. (2013) Antimicrobial nanostructures in food packaging. *Trends in Food Science and Technology*, **30**, 56-69.
80. Johnson, R. R., Johnson, A. T. C., and Klein, M. L. (2008) Probing the structure of DNA-carbon nanotube hybrids with molecular dynamics. *Nano Letters*, **8**, 69-75.
81. Hong Y., and Brown, D. G. (2008) Electrostatic behavior of the charge-regulated bacterial cell surface. *Langmuir*, **24**, 5003-5009.
82. Zuo, P., Li, X., Dominguez, D. C., and Ye, B. C. (2013) A PDMS/paper/glass hybrid microfluidic biochip integrated with aptamer-functionalized graphene oxide nano-biosensors for one-step multiplexed pathogen detection. *Lab on a Chip*, **13**, 3921-3928.
83. Yoo, S. M., Kim, D. K., and Lee, S. Y. (2015) Aptamer-functionalized localized surface plasmon resonance sensor for the multiplexed detection of different bacterial species. *Talanta*, **132**, 112-117.
84. Wu W. H., Li, M., Wang, Y., Ouyang, H. X., Wang, L., Li, C. X., Cao, Y. C., Meng, Q. H., and Lu, J. X. (2012) Aptasensors for rapid detection of Escherichia coli O157: H7 and Salmonella typhimurium. *Nanoscale Research Letters*, **7**, 658.
85. Chen, J., Jackson, A. A., Rotello, V. M., and Nugen, S. R. (2016) Colorimetric Detection of Escherichia coli Based on the Enzyme-Induced Metallization of Gold Nanorods. *Small*, **12**, 2469-2475.
86. Richards, S. J., Fullam, E., Besra, G. S., and Gibson, M. I. (2014) Discrimination between bacterial phenotypes using glyco-nanoparticles and the impact of polymer coating on detection readouts. *Journal of Materials Chemistry B*, **2**, 1490-1498.
87. Setterington, E. B., and Alocilja, E. C. (2011) Rapid electrochemical detection of polyaniline-labeled Escherichia coli O157:H7. *Biosensors & Bioelectronics*, **26**, 2208-2214.
88. Kang, X., Pang, G., Chen, Q., and Liang, X. (2013) Fabrication of Bacillus cereus electrochemical immunosensor based on double-layer gold nanoparticles and chitosan. *Sensors and Actuators B: Chemical*, **177**, 1010-1016.
89. Wu W., *et al.*, (2012) An aptamer-based biosensor for colorimetric detection of Escherichia coli O157:H7. *PLoS One*, **7**(11), e48999.
90. Li, B., Yu, Q., and Duan, Y. (2015) Fluorescent labels in biosensors for pathogen detection. *Critical Reviews in Biotechnology*, **35**, 82-93.
91. Liao, Y., Zhou, X., and Xing, D. (2014) Quantum dots and graphene oxide fluorescent switch based multivariate testing strategy for reliable detection of listeria monocytogenes. *ACS Applied Materials & Interfaces*, **6**, 9988-9996.
92. Zuo, P., Lu, X., Sun, Z., Guo, Y., and He, H. (2016) A review on syntheses, properties, characterization and bioanalytical applications of

fluorescent carbon dots. *Microchimica Acta*, **183**, 519-542.

93. Liu, J., Wagan, S., Dávila Morris, M., Taylor, J., and White, R. J. (2014) Achieving reproducible performance of electrochemical, folding aptamer-based sensors on microelectrodes: Challenges and prospects. *Analytical Chemistry*, **86**, 11417-11424.

94. Reddy, K. V. R., Yedery, R. D., and Aranha, C. (2004) Antimicrobial peptides: Premises and promises. *International Journal of Antimicrobial Agents*, **24**, 536-547.

95. Brogden, K. A. (2005) Antimicrobial peptides: Pore formers or metabolic inhibitors in bacteria? *Nature Reviews Microbiology*, **3**, 238-250.

96. Lupetti, A., Welling, M. M., Pauwels, E. K. J., and Nibbering, P. H. (2003) Radiolabelled antimicrobial peptides for infection detection. *Lancet Infectious Diseases*, **3**, 223-229.

97. Welling, M. M., Lupetti, A., Balter, H. S., Lanzzeri, S., Souto, B., Rey, A. M., Savio, E. O., Paulusma-Annema, A., Pauwels, E. K., and Nibbering, P. H. (2001) 99mTc-Labeled antimicrobial peptides for detection of bacterial and Candida albicans infections. *Journal of Nuclear Medicine*, **42**, 788-794.

98. Kulagina, N. V., Lassman, M. E., Ligler, F. S., and Taitt, C. R. (2005) Antimicrobial peptides for detection of bacteria in biosensor assays. *Analytical Chemistry*, **77**, 6504-6508.

99. Yoo, J. H., Woo, D. H., Chun, M. S., and Chang, M. S. (2014) Microfluidic based biosensing for Escherichia coli detection by embedding antimicrobial peptide-labeled beads. *Sensors and Actuators B: Chemical*, **191**, 211-218.

100. Kulagina, N. V., Shaffer, K. M., Anderson, G. P., Ligler, F. S., and Taitt, C. R. (2006) Antimicrobial peptide-based array for Escherichia coli and Salmonella screening. *Analytica Chimica Acta*, **575**, 9-15.

101. Mannoor, M. S., Zhang, S., Link, A. J., and McAlpine, M. C. (2010) Electrical detection of pathogenic bacteria via immobilized antimicrobial peptides. *Proceedings of the National Academy of Sciences of the United States of America*, **107**, 19207-19212.

102. Friedrich, C., Scott, M. G., Karunaratne, N., Yan, H., and Hancock, R. E. W. (1999) Salt-resistant alpha-helical cationic antimicrobial peptides. *Antimicrobial Agents and Chemotherapy*, **43**, 1542-1548.

103. Rydlo, T., Rotem, S., and Mor, A. (2006) Antibacterial properties of dermaseptin S4 derivatives under extreme incubation conditions. *Antimicrobial Agents and Chemotherapy*, **50**, 490-497.

104. Xu D., Jiang, L., Singh, A., Dustin, D., Yang, M., Liu, L., Lund, R., Sellati, T. J., and Dong, H. (2015) Designed supramolecular filamentous peptides: Balance of nanostructure, cytotoxicity and antimicrobial activity. *Chemical Communications*, **51**, 1289-1292.

105. Yang, M., Xu, D., Jiang, L., Zhang, L., Dustin, D., Lund, R., Liu, L., and Dong, H. (2014) Filamentous supramolecular peptide-drug conjug-

ates as highly efficient drug delivery vehicles. *Chemical Communications*, **50**, 4827-4830.

106. Liu, X., Marrakchi, M., Xu, D., Dong, H., and Andreescu, S. (2016) Biosensors based on modularly designed synthetic peptides for recognition, detection and live/dead differentiation of pathogenic bacteria. *Biosensors and Bioelectronics*, **80**, 9-16.
107. Wang Z.-F., Cheng, S., Ge, S.-L., Wang, H., Wang, Q.-J., He, P.-G., and Fang, Y.-Z. (2012) Ultrasensitive detection of bacteria by microchip electrophoresis based on multiple-concentration approaches combining chitosan sweeping, field-amplified sample stacking, and reversed-field stacking. *Analytical Chemistry*, **84**, 1687-1694.
108. Alhogail, S., Suaifan, G. A. R. Y., and Zourob, M. (2016) Rapid colorimetric sensing platform for the detection of Listeria monocytogenes foodborne pathogen. *Biosensors and Bioelectronics*, **86**, 1061-1066.
109. Jokerst, J. C., Adkins, J. A., Bisha, B., Mentele, M. M., Goodridge, L. D., and Henry, C. S. (2012) Development of a paper-based analytical device for colorimetric detection of select foodborne pathogens. *Analytical Chemistry*, **84**, 2900-2907.
110. Adkins, J. A., Boehle, K., Friend, C., Chamberlain, B., Bisha, B., and Henry, C. S. (2017) Colorimetric and electrochemical bacteria detection using printed paper- and transparency-based analytic devices. *Analytical Chemistry*, **89**, 3613-3621.
111. Gross, B. C., Erkal, J. L., Lockwood, S. Y., Chen, C., and Spence, D. M. (2014) Evaluation of 3D printing and its potential impact on biotechnology and the chemical sciences. *Analytical Chemistry*, **86**, 3240-3253.
112. Mandon, C. A., Blum, L. J., and Marquette, C. A. (2016) Adding biomolecular recognition capability to 3D printed objects. *Analytical Chemistry*, **88**(21), 10767-10772.
113. Ho, C. M. B., Ng, S. H., Li, K. H. H., and Yoon, Y. J. (2015) 3D printed microfluidics for biological applications. *Lab on a Chip*, **15**, 3727-3637.
114. Hribar, K. C., Soman, P., Warner, J., Chung, P., and Chen, S. (2014) Light-assisted direct-write of 3D functional biomaterials. *Lab on a Chip*, **14**, 268-275.
115. Spilstead, K. B., Learey, J. J., Doeven, E. H., Barbante, G. J., Mohr, S., Barnett, N. W., Terry, J. M., Hall, R. M., and Francis, P. S. (2014) 3D-printed and CNC milled flow-cells for chemiluminescence detection. *Talanta*, **126**, 110-115.
116. Gao, B., Yang, Q., Zhao, X., Jin, G., Ma, Y., and Xu, F. (2016) 4D Bioprinting for biomedical applications. *Trends in Biotechnology*, **34**, 746-756.
117. Bakarich, S. E., Gorkin, R., In Het Panhuis, M., and Spinks, G. M. (2015) 4D printing with mechanically robust, thermally actuating hydrogels. *Macromolecular Rapid Communications*, **36**, 1211-1217.

118. Klein, F., Richter, B., Striebel, T., Franz, C. M., von Freymann, G., Wgener, M., and Bastmeyer, M. (2011) Two-component polymer scaffolds for controlled three-dimensional cell culture. *Advanced Materials*, **23**, 1341-1345.

4

Recent Developments in Nanomaterials based Diagnostics, Targeted Drug Delivery, their Efficiency and Potential

Shadab Ahmed,* Naeem Shaikh, Nachiket Pathak and Akshay Sonawane

Institute of Bioinformatics and Biotechnology, Savitribai Phule Pune University (formerly University of Pune), Pune, India

Corresponding author: sahmed@unipune.ac.in

4.1 Introduction

Recent advances in the nanomaterial based molecular recognition have been discussed in this chapter, and the recognition may be for an enzyme, DNA or RNA, a specific antibody, or an aptamer, etc. [1,2]. The biomedical diagnostic applications based on optical, electrical, and electrochemical properties mainly rely on nanostructures for enhanced efficacy and performance, as is the case in bioelectronics devices [2]. Drug delivery using various carriers such as biomaterials has come a long way since its inception and has gained considerable attention from the biomedical, clinical and medical professionals as well as pharmaceutical companies. The search for the most efficacious and safe drug delivery system to improve the performance of the commonly used medicines available for human consumption will have significant consequence on the development of targeted anticancer drug delivery, peptide/protein delivery and gene therapy [2-4]. In this chapter, we have focused mainly on nanomaterials as biomaterials for targeted drug delivery, 3D nanofabrication methods and virus-based nanoparticles for biomedical applications.

4.2 Nanomaterials based Targeted Drug Delivery and their Potential

Drug targeting to the specific site inside the body is the main area of

Recent Trends in Nanobiotechnology, edited by Prakash Saudagar and K. Divakar
© 2019 Central West Publishing, Australia

interest for therapeutic research (Figure 4.1). Targeting drug to the specific site gives better efficiency of the drug and it reduces the damage to non-targeted cells in the body. Biological means for drug delivery, such as antibody mediated drug delivery, seem to be inadequate because of non-specific presence of antigenic determinants on the non-targeted cells [5]. Nanoparticles because of their size and other physio-chemical properties show interesting usage in vivo for drug delivery [5]. They can go through various body compartments and can reach not only the specific tissues but also specific cells and cellular structures. These properties of nanoparticles make them attractive for the research in the field of targeted drug delivery approaches [6]. Nanotechnology based drug delivery provides better efficacy by enhancing drug permeation, controlled release and targeting. Target specificity many a times plays a major role in many therapies such as anti-cancer chemotherapy [7]. Toxicity to the normal cells along with the neoplastic cells is the main drawback of chemotherapy. Imparting nanoparticles selectively increases the drug concentration in cancer cells, thus, avoiding the cytotoxicity to the normal cells [8].

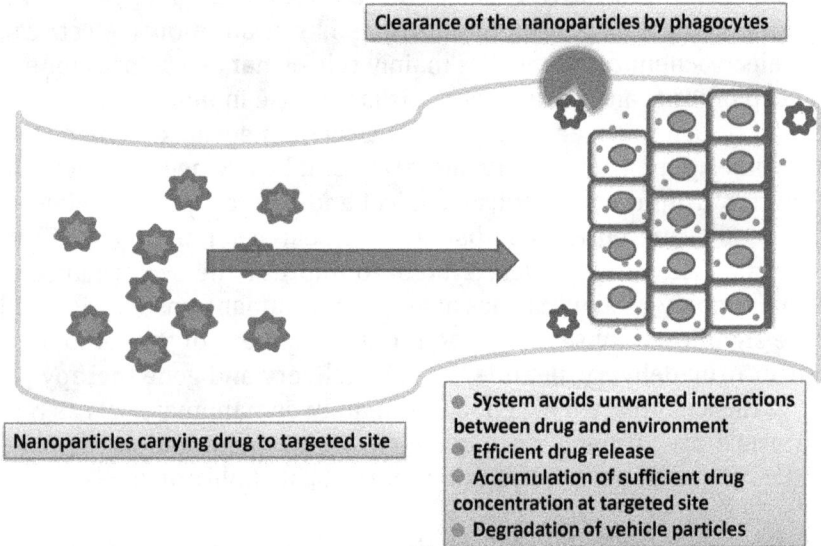

Figure 4.1 Schematic representation of nanoparticle mediated drug delivery system.

Release of the drug in the targeted cells is much enhanced by internalization of the nanosystem. As a result, it gets released in larger quantities and gives better therapeutic results [9]. Development of biocompatible and biodegradable nanoparticles have made it possible to develop drug delivery systems, such as albumin, lipids and chitosan [2,10,11]. As mentioned earlier, target selectivity is the main prerequisite in the case of anti-cancer drugs. Livatag®, doxorubicin loaded polyalkcyanoacrylate nanoparticles, developed by Bioalliance, is currently in its phase III clinical trial [12]. The drug is a major treatment regimen for the multidrug resistant hepatocarcinoma. Abraxane® is already being marketed, which is albumin-bound Paclitaxel for the treatment of metastatic breast cancer [13]. Despite significant research in the field of nanoparticles for drug delivery, a few materials have reached clinical trials and subsequent market for actual therapeutic use [13]. The possible reasons for this are given by Couvreur [13] as:

a) Less amount of drug is getting loaded because of the lesser percentage of the actual drug part than the carrier part or the high amount of carrier material is giving undesirable side effects.
b) Too rapid release of the drug after the administration with the carrier material. Mere adsorption of the drug particles on the surface of nanoparticles lead to their release before they reach the actual target.

Better drug loading and controlled release of drug can be achieved by development of nano-metal-oxide frameworks, i.e. nanoMOFs [14]. Composite nanospheres of metal oxides, such as $SiO_2@Fe/SiO_2$ composite spheres, exhibit dual functionality in cancer treatment. They can act as hyperthermic agent as well as a drug carrier [15]. Multifunctional nanoparticle construction includes the incorporation of drug and an imaging agent into the same nanoparticle. This enables simultaneous treatment and diagnosis as a part of personalized treatment of the patient. Theragnostics is a term representing such applications [16]. Nanoparticle mediated drug delivery has enabled pharmacology with certain capabilities such as controlled release, diffusion of the drug to the sites inaccessible for the drug, targeted delivery, improved drug metabolism and biodegradation .

In spite of the above-mentioned advancements, this area needs more research efforts in the direction of achieving improvement in the compatibility of the nanocarrier entities with the drug molecules and biological system.

4.3 Three dimensional or 3D Nanofabrication for Biomedical Applications

Nanofabrication is generally referred to as the process of fabrication of structures at nanometric scale i.e. having the dimensions of 10^{-9} m. Multi-layered fabrication leads to the generation of three dimensional structures which can have a number of applications from electronic circuit designing for drug delivery, bio-sensing and diagnostics such as detection of pathogens and parasites [17-19], cancer detection [20-26], detection of various biomarkers [28-31], etc. By coupling these fabricated structures with biomolecules, they can be used as nano-biosensors for disease detection, disease modelling, tissue engineering, etc.

4.3.1 Fabrication Methods

There are quite a few developed fabrication methods that are currently in use for biomedical applications. The various fabrication methods have been illustrated in Table 4.1 and in subsequent sections.

Soft lithography

Soft Lithography is a traditional technique which generally utilizes polymeric elastomers to create casts and molds to generate structures on appropriate substrates. One of the most commonly used polymers for this purpose is polydimethylsiloxane (PDMS) due to its high elasticity and minimal reactivity. The process consists of two steps:

a) Mold fabrication
b) Utilizing the mold to release the required pattern on the substrate.

The method can be used for a large range of dimensions, along with low cost.

Electrospinning

Electrospinning, as the name suggests, is a technique in which polymeric fibers are 'spun' using high voltage. Materials such as cellulose, chitin, dextran, etc., are widely used to make scaffolds, nanofiber-based biosensors, etc. The assembly generally consists of a syringe, conductive substrate and high voltage source. The resolution ranges from a few nanometers to a few micrometers.

Electron-beam Lithography

It is powerful technique similar to photolithography, but has a much higher resolution. The sample to be processed is coated with a layer of electron sensitive resist film, and the pattern is generated on it using a directed beam of electrons. Nanostructures like circuits, arrays, etc., can be fabricated using this method. This method can produce structures which have dimensions lower than 5 nm.

Atomic Force Deposition

Atomic Force Deposition method is used to directly deposit atoms or small complex molecules on a substrate in a controlled manner to generate structures such as arrays or scaffolds. This method can be used for the deposition of metals, small carbohydrates, etc., which can act as sensory molecules.

Table 4.1 Fabrication strategies employed for nanofabrication of biosensors (modified from [32])

Fabrication method	Dimension	Applications	Cost
Soft Lithography	5nm-100mm	arrays of nano-antennas, nano-circuits, graphene coated nanoparticles, inter-digited transducers, etc.	Low
Electrospinning	3nm-5μm	scaffolds, nanofiber-based biosensors, tissue engineering, nano-syringes, etc.	Low, but low yield
E-beam Lithography	Up to 5nm	Nano-arrays, nano-circuits, patterns, nano-holes, etc.	High
Atomic Force Deposition	Up to 5nm	Patterns, nanoparticle coatings, etc.	High

4.3.2 Device Structures

3D nanofabricated biosensors can be fabricated in a number of ways.

We will discuss some designs in detail in the following sections. A few more designs have been described in Table 4.2.

Arrays

A number of 3D fabrication methods utilize arrays of nanostructures for sensing of cells, biological molecules, etc. [33]. Dipalo *et al.* [34] fabricated nano-antenna arrays using nano-lithography to study the growth of neuronal cells and their activity to some extent. A similar needle-electrode microarray was developed by Lee *et al.* [30] for continuous monitoring of glucose levels in the blood. In another approach, Wu *et al.* [27] fabricated an array of 3D pillars of nickel and conjugated with antibodies for the detection of circulating tumorigenic cells (for cancer detection).

In a different approach, Gomez-Cruz *et al.* [35] created an array of nanoholes on a metallic surface and coupled with antibodies specific to the uropathogen *E. coli*. The detection was performed by monitoring the RI changes occurring after binding of the pathogen to these antibodies in real-time. Similarly, Soler *et al.* [36] used this method for real-time detection of pathogens causing ST diseases in the bodily fluids.

Nanoparticle-based

Nanomaterials such as SiO_2 nanoparticles, gold nanoparticles, carbon nanotubes, etc., have gained a large interest in recent years due to their very small size, high functionalization ability and electrical properties for use as biosensors. They can be coated with biosensitive materials, like antibodies, for use in diagnostics. Myung *et al.* [37] developed graphene coated nanoparticles and placed them between two Au electrodes. These were conjugated with antibodies specific to cancer cell surface markers for use in the detection of cancer cells. A similar method utilizing the electrical properties of nanoparticles was suggested by Lei *et al.* [32], which involved suspending the conjugated nanoparticles in a nanochannel fabricated in PDMS by soft lithography and monitoring the change in the conductance using platinum electrodes [31].

Other approaches focus on the optical properties of nanoparticles and couple them with Raman spectroscopy for analysis. This can be achieved by assembling smooth metal films made up of nanoparticles or by generating more complex nanostructures [20,37].

Nano-circuits, Patterns and Transducers

Nano-electronic circuits such as transistors or interdigited transducers are being widely applied for the detection of a number of biomolecules and cells. These are generally label free methods of detection. Bao *et al.* [20] demonstrated the fabrication of a nano-ribbon field effect transistor (FET) on a silicon substrate for the early detection of cancer using the carcinoembryonic antigen. It functioned by detecting the amperometric changes occurring over the FET after binding of the CEA antigen to its antibody conjugated on the surface of the fabricated transistor. Carbon nanostructures fabricated by nanolithography technique can be used in a similar fashion for the detection of a number of biological and chemical species depending on the molecule coupled to the sensor [38]. Interdigitated transducers or IDTs are widely employed in the fabrication of surface acoustic wave-based sensors. These are fabricated using different forms of nanolithography and can be used for a number of bio-sensing applications such as detection of antibodies, antigens, toxins, cancer cells, bodily fluids, etc.

Another type of pattern consists of nano-needles, where two nanoelectrodes are separated by an insulating material and conjugated to a biomolecule such as an aptamer or an antibody. By monitoring the conductivity changes occurring between these electrodes, one can detect the presence of a specific molecule or cell in the sample [29]. Nanowires, fabricated using methods such as nanolithography and deposition techniques, are also used for bio-sensing. Antibodies can be conjugated on the surface of these nanowires and can function in a similar manner as other electrical biosensors [28].

Table 4.2 Alternative strategies and materials utilized for development or fabrication of high performance biosensors

Method	Materials	Applications
Paper or thread based	Cellulose, nitrocellulose, cotton	Paper based ELISA, colorimetric assays, immune-assays, etc.
Scaffold type	Silk, chitosan, cellulose, DNA	Implants, cell growth monitoring, drug delivery, etc.
Carbon Nanotube based	Carbon nanotube electrodes	Cancer diagnostics, cell surface marker identification, etc.
Magnetic Nanoparticles	Polymer nanoparticles coated with iron oxide	Tumor imaging by magnetic resonance imaging (MRI)

4.3.3 Biomedical Applications of 3D Fabricated Nanobiosensors

Nanofabricated structures are being widely used for a number of molecular and biomedical applications (Figure 4.2). These enable fast, cheap and Point of Care (PoC) detection methods. These are generally non-invasive and require minimal sample for detection. Most of these devices can be included in the ASSURED (Affordable, Sensitive, Specific, User-friendly, Rapid and Robust, Equipment-free and Deliverable to end-users) criteria of the World Health Organization [39]. Analysis of cell growth and proliferation [32,36], apoptosis [40], differentiation [41,42] and movements [43] can be carried out on such nanofabricated devices. Detection of tumorigenic or cancer cells has been a major area of focus for these nanobiosensors, with a number of applications including nanoparticles for cancer cell identification [36,44,45], arrays of nanostructures [27], SAW based devices [21,26,46] and carbon nanostructures such as nanotubes [24] as well as other techniques discussed above have been employed for detection of cancer cells using cell surface antigens. Detection of pathogens

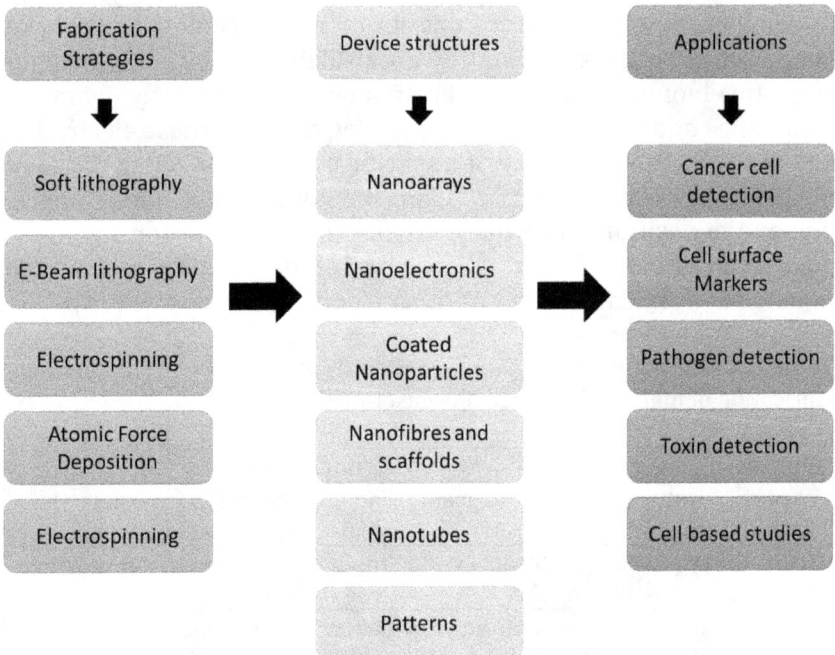

Figure 4.2 Overview of strategies for three-dimensional nanofabrication and their biosensing applications.

such as the uropathogen *E. coli* [35] as well as pathogens causing ST diseases such as *Chlamydia trachomatis* and *Neisseria gonorrhoea* [36] can be achieved directly from the urine samples by the use of these biosensors. Another approach for the detection of bacterial pathogens is by coupling the nanostructures with SPR. This can be used for the detection and identification of a number of bacterial species accurately [47]. A number of biomolecules such as proteins, sugars, nucleotides and even strands of DNA can be detected accurately by using nanowire and nanoarrays based devices. This can help in real-time estimation of blood glucose levels for patients with diabetes.

4.3.4 Advantages over Traditional Diagnostic Tools

3D fabricated nanobiosensors are cost-effective as most sensors are fabricated using silicon substrates, polymers, nanoparticles or a combination of these. These materials do not require large sophisticated fabrication units. They have very low to no power requirements, thus, can be used in fields and remote areas. These are generally simple to use and robust, thus, reduce the need for trained professionals. As the scale of the sensing apparatus is nanometric, which is comparable to the size of molecules and cells, these units have high sensitivity and specificity. Most of the devices are multifunctional, i.e. by simply changing the biomolecule coupled to the nanostructures, a wide range of analytes can be detected by a single device. The sensing is reliable and requires small sample volumes in most cases.

4.4 Advances in Applications of Virus-based Nanoparticles for Biomedical Applications

Virus-based nanoparticles can be defined as highly defined structurally modular three-dimensional structures based on proteins derived from viruses. Thus, the use of virus-based nanoparticles can be regarded as an equivalent of 3D nanotechnology methods [48]. Virus-based nanoparticles are known to have advantages over other available nanoparticles as they are biocompatible, biodegradable, easy to functionalize and provide accurate delivery of cargo to the target with the ability to cross most cell membranes. The size range, starting from a few microns to around 10 nm, with a variety of shapes, usually icosahedral, also provide them an upper hand over traditional nanoparticles. Even though bacteriophage and plant viruses are regarded as safe for humans, there is a risk that the use of virus-based nano-particles for therapeutics may result in some pathogenic response due to host-virus response [49,50].

Throughout history, viruses have been known to cause infections at a global level, and it has been estimated that around 15% of cancer deaths are due to some sort of viral infection. However, at times, we have employed viruses to beat various diseases like smallpox, polio, etc., by vaccination [51]. Viruses contain several protein molecules which self-assemble for packing in the viral nucleic acid. Advances in genetic engineering have shown us a path to genetically modify viruses and use them for various purposes, with one such purpose is to use them as scaffolds [49]. In 1998, Trevor Douglas and Mark Young for the first time demonstrated that viral-based self-assembly coat proteins can be used for the preparation of nanometer structures. They showed that viral coat proteins of cowpea chlorotic mottle virus (CCMV) can be synthetically modulated to be used as drug transport-ers and for mineralization purposes [52]. Since then, viruses have provided an ideal starting point for the synthesis of homogeneous na-noparticles as an alternative to current methods for large-scale syn-thesis of structurally homogeneous nanoparticles. Currently, with all the advancements and upcoming techniques, virus-based nanoparti-cle systems present us with an opportunity to develop various bio-medical applications such as imaging, drug delivery, vaccines against infectious diseases, scaffolding, etc.

4.4.1 Advent of Virus-based Nanoparticles for Therapeutic Applications

One of the hurdles in therapeutics is the efficiency and target speci-ficity. Various techniques have been employed for target delivery to minimize side effects. Viral capsomeres contain pores that allow dif-fusion of small molecules through them, molecules can be drugs, met-als or hormones. Treatments such as low pH, electrostatic interac-tions and surface charge modification can help in the retention of these molecules [53]. Cowpea mosaic virus (CPMV) and flock house virus (FHV) based nanoparticles have been proven to be good con-tenders for drug delivery, gene therapy and diagnostic purposes [54]. In this respect, some of these are in the clinical trial phases to be used as oncotherapy [51]. Herein, in the below sections, we focus on the drug delivery and imaging-based application of virus-based nanopar-ticles.

4.4.2 Drug Delivery using Virus-based Nanoparticles

Protein cages formed by viruses present characteristic features of size uniformity and self-assembly. A given set of viral coat particles can form structures of same structural properties n number of times with minimum deviation. This property is quite important when we propose the use of viral nanoparticles for therapeutic uses as it contributes to the drug carrying capacity and efficiency of the nanoparticles. Development of target-specific viral nanoparticles has allowed us to deliver toxic payloads to targeted cells without damaging healthy cells. This can be easily achieved by either encapsulation or conjugation of compounds with viral coat proteins. Toxic cargos are preferentially loaded inside the capsid rather than coating the surface, the viral coat here also provides protection from enzymatic degradation and unintentional interactions with other compounds which may give rise to cross-reactivity. This also provides us with control over drug distribution profile and monitoring.

Zeng *et al.* [55] showed that cucumber mosaic virus (CMV) coat protein-based encapsulation can be used to carry doxorubicin, an anthracycline antibiotic. Further treatment of CMV capsid protein with folic acid showed an increase in the amount of doxorubicin encapsulation. Results have shown that encapsulation of doxorubicin enhances anti-tumor properties in rats and reduces doxorubicin-induced cardiotoxicity [55]. Amalgamation of polyethylene glycol polymer (PEG) has shown to inhibit anti-viral nanoparticle responses in mice. PEGylation also inhibits internalization of nanoparticles by cells, increases plasma circulation and also enhances shielding efficiency. Increase in chain length of PEG was shown to enhance the efficiency of the nanoparticles [56]. Cowpea chlorotic mottle virus (CCMV) is also a popular candidate for virus-based nanoparticle synthesis and is one of the first few viruses to be used for the synthesis of nanoparticles. CCMV does not have surface-active cysteine, and it has been genetically modified to present surface active cysteine for binding of a variety of molecules, such as fluorophores, biotin, PEG polymers, organic dyes, polypeptides, antibodies, etc. [57].

The capacity of viral nanoparticles has been shown to differ from species to species. For example, around 300 doxorubicin molecules can be loaded on the surface of cowpea mosaic virus (CPMV), while hibiscus chlorotic mosaic virus (HCRSV) shows the potential to be loaded with around 950 molecules of doxorubicin [58,59]. This encapsulation capacity depends upon the availability of surface-active groups on coat protein surface and treatments such as folic acid conjugation or PEGylation on the coat surface.

Virus coat proteins can be modified to form cargo encapsulations according to our needs based on the following principles:

a) Binding properties of self-assembling viral coat proteins [20].
b) Surface charge and electrostatic interactions between the capsomeres [60].
c) Activating amino acid side chains such as the carboxylate group of glutamate and aspartate, sulfhydryl side chain of cysteine or phenol groups of tyrosine residues [61].
d) Genetic techniques such as modification of the viral structural genes [62].

Some of the viruses that have been used for cargo delivery have been illustrated in Table 4.3, which confirms considerable improvement in targeted delivery and efficacy.

Table 4.3 Important viral delivery systems and cargo

Virus type	Virus name	Cargo	Ref.
Bacterio-phages	Acinetobacter phage AP205	protein	[29]
	Enterobacteria phage P22	fluorophore	[41]
	Enterobacteria phage λ	fluorophore, protein	[19]
	HK97	fluorophore, protein	[16]
	M13	fluorophore	[13]
	MS2	oligonucleotide, protein	[42]
	Qβ	fluorophore, oligonucle-otide, protein	[3,32]
Plant Viruses	Brome mosaic virus (BMV)	fluorophore, protein	[46]
	Cowpea Chlorotic Mottle Virus (CCMV)	peptides, biotin, fluoro-phores	[14]
	Cowpea mosaic virus (CPMV)	stilbene, doxorubicin, fluorophore	[3]
	Cucumber mosaic virus (CMV)	Folic acid	[48]
	Hibiscus chlorotic ring-spot virus (HCRSV)	Folic acid	[33]

	Papaya Mosaic Virus (PapMV)	peptide	[4,34]
	Potato virus X (PVX)	heterologous peptide	[24]
	Red Clover Necrotic Mosaic Virus (RCNMV)	peptide	[38]
	Tobacco mosaic virus (TMV)	biotin, fluorophore, peptide	[6]
	Turnip yellow mosaic virus (TYMV)	fluorophore	[5]
Polyoma-virus	John Cunningham virus (JC virus)	fluorophore	[27]
	Murine polyomavirus (MPyV)	peptide	[36]
	Simian virus 40 (SV40)	protein	[40]
Others	Bovine papillomavirus type 1 (BPV1)	peptide	[31]
	Canarypox virus (CNPV)	peptide	[18]
	Hepatitis B virus (HBV)	protein	[21]
	Infectious Bursal Disease Virus (IBDV)		[28]
	Influenza	protein	[30]

4.4.3 Virus-based Imaging Probes

Imaging is an important tool in medicine for diagnostic and visualization purposes, with early disease detection and prognosis. Viruses are being used for tissue-specific imaging. They are used in magnetic resonance imaging (MRI), positron emission tomography (PET), fluorescence-based imaging, etc. The ease of usage, assembly, precise targeting and easy removal from the body are few benefits of using virus-based nanoparticles [63-70]. Table 4.4 demonstrates the viruses used for imaging probes.

Fluorescence imaging

Feng *et al.* [71] for the first time observed viral behavior in mammalian cells with the help of self-assembled capsid-quantum-dot hybrid particles. The quantum dots (QD) are also encapsulated by SV40 viral particles. PEGylation of virus nanoparticles has shown to elongate blood circulation time and reduced chances of accumulation in the reticuloendothelial system. Further, QDs were encapsulated inside the core of pseudotyped HIV-1-based lentivirus (PTLV) which was confirmed by single particle tracking of QD. Unmatched spatiotemporal resolution has been achieved by encapsulation of Ag_2S QDs in virus-based nanoparticles [72].

Table 4.4 Viruses used for delivery for imaging probes

Type	Name	Ref.
Bacteriophage	MS2	[64]
	Enterobacteria phage P22	[64]
	Cowpea Chlorotic Mottle Virus (CCMV)	[65]
	Cowpea mosaic virus (CPMV)	[65]
Plant Viruses	Brome mosaic virus (BMV)	[66]
	Red clover necrotic mosaic virus (RCNMV)	[65]
	Tomato mosaic virus (ToMV)	[65]
	Tobacco mosaic virus (TMV)	[65]
Polyomavirus	SV40 (simian virus 40)	[67]
	HBV (hepatitis B core virus-like particles)	[68]
	Alphavirus	[65]
Other	PTLV (pseudotyped HIV-1-based lentivirus	[69]
	HIV-1 (human immunodeficiency virus type 1)	[70]

Magnetic Resonance Imaging (MRI)

Superparamagnetic iron oxide nanoparticles are used for MRI purposes. Viral capsids enhance relaxation rates and provide an ideal platform for the modification of contrast labeling agents [73]. Douglas *et al.* [74] developed Gd^{3+} attached CCMV viral nanoparticles as MRI contrast agents. Iron oxide-based nanoparticles were also reported to exhibit better r2/r1 (transverse/longitudinal relativity) ratio compared to commercial contrast agents Feridex® and Supravist® [65].

4.5 Outlook

The recent discoveries and reports with regards to targeted drug delivery and diagnostics hold a lot of promise for the development of nanomaterial-based carrier systems with better efficacy and less toxicity. Also, the development of new methods and materials signifies that more drugs can now be delivered in a targeted fashion, along with increased efficacy to fight critical diseases/disorders. Many of the nanomaterial-based systems have been commercialized with considerable success in recent times, and this has generated a renewed interest in the pharmaceutical market for research in this direction. Also, the investment in the research on drug delivery and diagnostics has been increased in many countries to develop systems to improve efficacy of all commonly used drugs. With the advent of 3D printing and 3D fabrication methods, new doors towards creating newer and efficient integrated circuits with advance photonics have been opened, which may lead to even higher sensitivity in upcoming systems. However, control on the mechanical properties, material porosity and biocompatibility by way of further functionalization is needed for achieving biodegradability and therapeutic drug-delivery. There is a significant expectation that these hurdles will be overcome in near future.

Acknowledgments

The authors are grateful to Savitribai Phule Pune University (SPPU) and to an extent to University Grants Commission (UGC) New Delhi for providing the necessary infrastructure.

References

1. Oliveira, O. N., Iost, R. M., Siqueira, J. R. Crespilho, F. N. and Caseli, L. (2014) Nanomaterials for diagnosis: challenges and applications in smart devices based on molecular recognition. *ACS Applied Materials and Interfaces*, **6**(17), 14745-14766.
2. Suri, S. S, Fenniri, H., and Singh, B. (2007) Nanotechnology-based drug delivery systems. *Journal of Occupational Medicine and Toxicology*, **2**, 1-16.
3. Parveen, S., Misra, R., and Sahoo, S. K. (2012) Nanoparticles: a boon to drug delivery, therapeutics, diagnostics and imaging. *Nanomedicine: Nanotechnology, Biology and Medicine*, **8**(2), 147-166.
4. Patra, J. K., Das, G., Fraceto, L. F., Campos, E. V. R., Rodriguez- Torres, M. d. P., Acosta-Torres, L. S., Diaz-Torres, L. A., Grillo, R., Swamy, M. K., Sharma, S. Habtemariam, S., and Shin, H.-S. (2018) Nano based drug delivery systems:recent developments and future prospects. *Journal of Nanobiotechnolgy*, **16**(71), 1-33.
5. Davis, S. S. (1997) Biomedical applications of nanotechnology-implications for drug targeting and gene therapy. *Trends in Biotechnology*, **15**(6), 217-224.

6. Li, Y., Xiao, K., Zhu, W., Deng, W., and Lam K. S. (2014) Stimuli-responsive cross-linked micelles for on-demand drug delivery against cancers. *Advanced Drug Delivery Reviews,* **66**, 58-73.

7. Viswanathan, S. S., Choe, H. C., and Yeung, K. W. K. (2011) Nanotechnology in biomedical applications: a review. *International Journal of Nano and Biomaterials,* **3**(2), 119-139.

8. Allen, T. M. (2002) Ligand-targeted therapeutics in anticancer therapy. *Nature Reviews Cancer,* **2**(10), 750-763.

9. Leucuta, S. E. (2010) Nanotechnology for delivery of drugs and biomedical applications. *Current Clinical Pharmacology,* **5**(4), 257-280.

10. Marty, J. J. (1978) Nanoparticles-a new colloidal drug delivery system. *Pharmaceutica Acta Helvetiae,* **53**, 17-23.

11. Siekmann, B., and Westesen, K. (1992) Submicron-sized parenteral carrier systems based on solid lipids. *Pharmaceutics and Pharmacology Letters,* **1**(3), 123-126.

12. Calvo, P., Remuñán-López, C., Vila-Jato, J. L., and Alonso, M. J. (1997) Novel hydrophilic chitosan-polyethylene oxide nanoparticles as protein carriers. *Journal of Applied Polymer Science,* **63**(1), 125-132.

13. Couvreur, P. (2013) Nanoparticles in drug delivery: past, present and future. *Advanced Drug Delivery Reviews,* **65**(1), 21-23.

14. Horcajada, P., Chalati, T., Serre, C., Gillet, B., Sebrie, C., Baati, T., Eubank, J. F., Heurtaux, D., Clayette, P., Kreuz, C., Chang, J.-S., Hwang, Y. K., Marsaud, V., Bories, P.-N., Cynober, L., Gil, S., Férey, G., Couvreur, P. and Gref, R. (2010) Porous metal-organic-framework nanoscale carriers as a potential platform for drug delivery and imaging. *Nature Materials,* **9**(2), 172-178.

15. Pin-Wei, H., Tseng, C. L., and Kuo, D. H. (2015) Preparation of SiO_2-protecting metallic Fe nanoparticle/SiO_2 composite spheres for biomedical application. *Materials,* **8**(11), 7691-7701.

16. Arias, J. L., Reddy, L. H., Othman, M., Gillet, B., Desmaële, D., Zouhiri, F., Dosio, F., Gref, R., and Couvreur, P. (2011) Squalene based nanocomposites: a new platform for the design of multifunctional pharmaceutical theragnostics. *ACS Nano,* **5**(2), 1513-1521.

17. Jafari, H., Amiri, M., Abdi, E., Navid, S. L., Bouckaert, J., Jijie, R., Boukherroub, R., and Szunerits, S. (2019) Entrapment of uropathogenic E. coli cells into ultra-thin sol-gel matrices on gold thin films: A low cost alternative for impedimetric bacteria sensing. *Bisensors and Bioelectronics,* **124-125**, 161-166.

18. Tigli, O., Bivona, L., Berg, P., and Zaghloul, M. E. (2010) Fabrication and characterization of a surface-acoustic-wave biosensor in CMOS technology for cancer biomarker detection. *IEEE Transactions on Biomedical Circuits and Systems,* **4**(1), 62-73.

19. Majdinasab, M., Hayat, A., and Marty, J. L. (2018) Aptamer-based assays and aptasensors for detection of pathogenic bacteria in food samples. *Trends in Analytical Chemistry,* **107**, 60-77.

20. Bao, Z., Sun, J., Zhao, X., Li, Z., Cui, S., Meng, Q., Zhang, Y., Wang, T., and Jiang, Y. (2017) Top-down nanofabrication of silicon nanoribbon field effect transistor (Si-NR FET) for carcinoembryonic antigen detection. *International Journal of Nanomedicine*, **12**, 4623-4631.
21. Chang, K., Pi, Y., Lu, W., Wang, F., Pan, F., Li, F., Jia, S., Shi, J., Deng, S., and Chen, M. (2014) Label-free and high-sensitive detection of human breast cancer cells by aptamer-based leaky surface acoustic wave biosensor array. *Biosensors and Bioelectronics*, **60**, 318-324.
22. Dragonieri, S., Annema, J. T., Mar, R. S., van der Scheea, P. C., Spanevello, A., Carratú, P., Resta, O., Rabe, K. F.,and Sterk, P. J. (2009) An electronic nose in the discrimination of patients with non-small cell lung cancer and COPD. *Lung Cancer*, **64**(2), 166-170.
23. Zerrouki, C., Fourati, N., Lucas, R., Vergnaud, J., Fougnion, J. M., Zerrouki, R., and Pernelle, C. (2010) Biological investigation using a shear horizontal surface acoustic wave sensor: small "click generated" DNA hybridization detection. *Biosensors and Bioelectronics*, **26**, 1759-1762.
24. Ji, S. R., Liu, C., Zhang, B., Yang, F., Xu, J., Long, J., Jin, C., Fu, D.-l., Ni, Q.-x., and Yu, X.-j. (2010) Carbon nanotubes in cancer diagnosis and therapy. *Biochimica et Biophysica Acta* (BBA) - *Reviews on Cancer*, **1806**(1), 29-35.
25. Ravalli, A., da Rocha, C. G., Yamanaka, H., and Marrazza, G. (2015) A label-free electrochemical affisensor for cancer marker detection: The case of HER2. *Bioelectrochemistry*, **106**, 268-275.
26. Gruhl, F. J., and Lange, K. (2012) Influence of surface preparation parameters on the signal response of an acoustic biosensor for the detection of a breast cancer marker. *IEEE Sensors Journal*, **12**, 1647-1648.
27. Wu, X., Xiao, T., Luo, Z., He, R., Cao, Y., Guo, Z., Zhang, W., and Chen, Y. (2018) A micro-/nano-chip and quantum dots-based 3D cytosensor for quantitative analysis of circulating tumor cells. *Journal of Nanobiotechnology*, **16**(1), 65.
28. Carrara, S., Sacchetto, D., Doucey, M.-A., Baj-Rossi, C., De Micheli, G., and Leblebici, Y., (2012) Memristive-biosensors: A new detection method by using nanofabricated memristors. *Sensors and Actuators B: Chemical*, **171**, 449-457.
29. Esfandyarpour, R., Javanmard, M., Koochak, Z., Harris, J. S., and Davis, R. W. (2014) Matrix Independent Label-free Manoelectronic Biosensor. *IEEE 27th International Conference on Micro Electro Mechanical Systems* (MEMS), USA, pp. 1083-1086.
30. Lee, S. J., Yoon, H. S., Xuan, X., Park, J. Y., Paik, S. J., and Allen, M. G. (2016) A patch type non-enzymatic biosensor based on 3D SUS micro-needle electrode array for minimally invasive continuous glucose monitoring. *Sensors and Actuators B: Chemical*, **222**, 1144-1151.

31. Fens, N., Zwinderman, A. H., van der Schee, M. P., de Nijs, S. B., Dijkers, E., Roldaan, A. C., Cheung, D., Bel, E. H., and Sterk, P. J. (2009) Exhaled breath profiling enables discrimination of chronic obstructive pulmonary disease and asthma. *American Journal of Respiratory and Critical Care Medicine*, **180**, 1076-1082.

32. Lei, Y., Xie, F., Wang, W., Wu, W., and Li, Z. (2010) Suspended nanoparticle crystal (S-NPC): A nanofluidics-based, electrical read-out biosensor. *Lab on a Chip*, **10**(18), 2338-2340.

33. Qian, T., and Wang, Y. (2010) Micro/nano-fabrication technologies for cell biology. *Medical and Biological Engineering and Computing*, **48**(10), 1023-1032.

34. Dipalo, M., Messina, G. C., Amin, H., La Rocca, R., Shalabaeva, V., Simi, A., Maccione, A., Zilio, P., Berdondini, L., and De Angelis, F. (2015) 3D plasmonic nanoantennas integrated with MEA biosensors. *Nanoscale*, **7**(8), 3703-3711.

35. Gomez-Cruz, J., Nair, S., Manjarrez-Hernandez, A., Parra, S. G., Ascanio, G., and Escobedo, C., (2018) Cost-effective flow-through nanohole array-based biosensing platform for the label-free detection of uropathogenic *E. coli* in real time. *Biosensors and Bioelectronics*, **106**, 105-110.

36. Soler, M., Belushkin, A., Cavallini, A., Kebbi-Beghdadi, C., Greub, G., and Altu, H. (2017) Multiplexed nanoplasmonic biosensor for one-step simultaneous detection of Chlamydia trachomatis and Neisseria gonorrhoeae in urine. *Biosensors and Bioelectronics*, **94**, 560-567.

37. Myung, S., Solanki, A., Kim, C., Park, J., Kim, K. S., and Lee, K. B. (2011) Graphene-encapsulated nanoparticle-based biosensor for the selective detection of cancer biomarkers. *Advanced Materials*, **23**(19), 2221-2225.

38. Banholzer, M. J., Millstone, J. E., Qin, L., and Mirkin, C. A. (2008) Rationally designed nanostructures for surface-enhanced Raman spectroscopy. *Chemical Society Reviews*, **37**(5), 885-897.

39. Lee, J. A., Lee, S. S., Lee, K. C., Park, S. I., Woo, B. C., and Lee, J.-O. (2007) Biosensor utilizing resist-derived carbon nanostructures. *Applied Physics Letters*, **90**(26), 264103.

40. Kosack, C. S., Page, A. L., and Klaster, P. R. (2017) A guide to aid the selection of diagnostic tests. *Bulletin of the World Health Organization*, **95**(9), 639.

41. Chen, C. S., Mrksich, M., Huang, S., Whitesides, G. M., and Ingber, D. E. (1997) Geometric control of cell life and death. *Science*, **276**(5317), 1425-1428.

42. McBeath, R., Pirone, D. M., Nelson, C. M., Bhadriraju, K., and Chen, C. S. (2004) Cell shape, cytoskeletal tension, and *RhoA* regulate stem cell lineage commitment. *Developmental Cell*, **6**(4), 483-495.

43. Ruiz, S. A., and Chen, C. S. (2008) Emergence of patterned stem cell

differentiation within multicellular structures. *Stem Cells*, **26**(11), 2921-2927.

44. Bailly, M., Yan, L., Whitesides, G. M., Condeelis, J. S., and Segall J. E. (1998) Regulation of protrusion shape and adhesion to the substratum during chemotactic responses of mammalian carcinoma cells. *Experimental Cell Research*, **241**(2), 285-299.

45. Lee, K. B., Myung, S., and Solanki, A. (2015) Graphene-encapsulated Nanoparticle-based Biosensor for the Selective Detection of Biomarkers, patent US9162885.

46. Gronewold, T. M., Baumgartner, A., Quandt, E., and Famulok, M. (2006) Discrimination of single mutations in cancer-related gene fragments with a surface acoustic wave sensor. *Analytical Chemistry*, **78**(14), 4865-4871.

47. Yoo, S. M., Kim, D.K., and Lee, S. Y. (2015) Aptamer-functionalized localized surface plasmon resonance sensor for the multiplexed detection of different bacterial species. *Talanta*, **132**, 112-117.

48. *Nanowrek Spotlight*. Online: https://www.nanowerk.com/spotlight/spotid=1635.php [accessed 26th October 2018].

49. Strable, E., and Finn, M. G. (2009) Chemical modifications of viruses and virus-like particles, viruses and nanotechnology. *Current Topics in Microbiology*, **327**, 1-21.

50. Koudelka, K. J., Pitek, A. S., Manchester, M., and Steinmetz, N. F. (2015) Virus-based nanoparticles as versatile nanomachines. *Annual Review of Virology*, **2**, 379-401.

51. Cattaneo, R., and Russell, S. J. (2017) How to develop viruses into anticancer weapons. *PLoS Pathogens*, **13**(3), e1006190.

52. Douglas, T., and Young, M. (1998) Host-guest encapsulation of materials by assembled virus protein cages. *Nature*, **393**(6681), 152-155.

53. Yildiz, I., Lee, K. L., Chen, K., Shukla, S., and Steinmetz, N. F. (2013) Infusion of imaging and therapeutic molecules into the plant virus-based carrier cowpea mosaic virus: Cargo-loading and delivery. *Journal of Controlled Release*, **172**(2), 568-578.

54. Destito, G., Schneemann, A., and Manchester, M. (2009) Biomedical nanotechnology using virus-based nanoparticles. *Current Topics in Microbiology and Immunology*, **327**, 95-122.

55. Zeng, Q., H Wen, Q Wen, X Chen, Y Wang, W Xuan, Liang, J., and Wana, S. (2013) Cucumber mosaic virus as drug delivery vehicle for doxorubicin. *Biomaterials*, **34**(19), 4632-4642.

56. Schoonen, L., and van Hest, J. C. (2014) Functionalization of protein-based nanocages for drug delivery applications. *Nanoscale*, **6**(13), 7124-7141.

57. Wen, A. M., Shukla, S., Saxena, P., Aljabali, A. A. A., Yildiz, I., Dey, S., Mealy, J. E., Yang, A. C., Evans, D. J., Lomonossoff, G. P., and Steinmetz, N. F. (2012) Interior engineering of a viral nanoparticle and its tum-

or homing properties. *Biomacromolecules*, **13**(12), 3990-4001.

58. Aljabali, A.A.A., Shukla, S., Lomonossoff, G. P., Steinmetz, N. F., and Evans, D. J. (2012) CPMV-Dox delivers. *Molecular Pharmaceutics*, **10**(1), 3-10.

59. Ren, Y., Wong, S. M., and Lim, L. Y. (2007) Folic acid-conjugated protein cages of a plant virus: a novel delivery platform for doxorubicin. *Bioconjugate Chemistry*, **18**(3), 836-843.

60. Daniel, M. C., Tsvetkova, I. B., Quinkert, Z. T., Murali, A., De, M., Rotello, V. M., Kao, C. C., and Dragnea, B. (2010) Role of surface charge density in nanoparticle-templated assembly of bromovirus protein cages. *ACS Nano*, **4**, 3853-3860.

61. Schlick, T. L., Ding, Z., Kovacs, E. W., and Francis M. B. (2005) Dual-surface modification of the tobacco mosaic virus. *Journal of the American Chemical Society*, **127**(11), 3718-3723.

62. Zeltins, A. (2015) Viral nanoparticles: Principles of construction and characterization. In: *Viral Nanotechnology*, Khudyakov, Y., and Pumpen, P. (eds.), CRC Press, USA, pp. 112-139.

63. Wen, A. M., and Steinmetz, N. F. (2016) Design of virus-based nano-materials for medicine, biotechnology, and energy. *Chemical Society Reviews*, **45**(15), 4074-4126.

64. Capehart, S. L., Coyle, M. P., Glasgow, J. E., and Francis M. B. (2013) Controlled integration of gold nanoparticles and organic fluoro-phores using synthetically modified MS2 viral capsids. *Journal of the American Chemical Society*, **135**(8), 3011-3016.

65. Zhang, W., Xu, C., Yin, G. Q., Zhang, X. E., Wang, Q., and Li, F. (2017) Encapsulation of inorganic nanomaterials inside virus-based nano-particles for bioimaging. *Nanotheranostics*, **1**(4), 358-368.

66. Huang, X., Stein, B. D., Cheng, H., Malyutin, A., Tsvetkova, I. B., Baxter, D. V., Remmes, N. B., Verchot, J., Kao, C., Bronstein, L. M., and Drag-nea, B. (2011) Magnetic virus-like nanoparticles in N. benthamiana plants: A new paradigm for environmental and agronomic biotech-nological research. *ACS Nano*, **5**(5), 4037-4045.

67. Sun, X., Li, W., Zhang, X., Qi, M., Zhang, Z., Zhang, X. E., and Cui, Z. (2016) In vivo targeting and imaging of atherosclerosis using mul-tifunctional virus-like particles of simian virus 40. *Nano Letters*, **16**(10), 6164-6171.

68. Malyutin, A. G., Easterday, R., Lozovyy, Y., Spilotros, A., Cheng, H., Sanchez-Felix, O. R., Stein, B. D., Morgan, D. G., Svergun, D. I., Drag-nea, B., and Bronstein, L. M. (2014) Virus like nanoparticles with maghemite cores allow for enhanced MRI contrast agents. *Chemis-try of Materials*, **27**(1), 327-335.

69. Zhang, Y., Ke, X., Zheng, Z., Zhang, C., Zhang, Z., Zhang, F., Hu, Q., He, Z., and Wang, H. (2013) Encapsulating quantum dots into enveloped virus in living cells for tracking virus infection. *ACS Nano*, **7**(5), 3896-3904.

70. Li, Q., Li, W., Yin, W., Guo, J., Zhang, Z. P., Zeng, D., Zhang, X., Wu, Y., Zhang, X.-E., and Cui, Z. (2017) Single-particle tracking of human immunodeficiency virus type 1 productive entry into human primary macrophages. *ACS Nano*, **11**(4), 3890-3903.

71. Feng, L., Zhang, Z. P., Peng, J., Cui, Z.Q., Pang, D. W., Li, K., Wei, H. P., Zhou, Y.-F., Wen, J.-K., and Zhang, X.-E. (2009) Imaging viral behavior in mammalian cells with self -assembled capsid-quantum-dot hybrid particles. *Small*, **5**(6), 718-726.

72. Li, C., Li, F., Zhang, Y., Zhang, W., Zhang, X.-E., and Wang, Q. (2015) Real-time monitoring surface chemistry-dependent in vivo behaviors of protein nanocages via encapsulating an NIR-II Ag2S quantum dot. *ACS Nano*, **9**(12), 12255-12263.

73. Liepold, L., S., Anderson, Willits, D., Oltrogge, L., Frank, J. A., Douglas, T., and Young*et* M. (2007) Viral capsids as MRI contrast agents. *Magnetic Resonance in Medicine: An Official Journal of the International Society for Magnetic Resonance in Medicine*, **58**(5), 871-879.

74. Allen, M., Bulte, J. W. M., Liepold, L., Basu, G., Zywicke, H. A., Frank, J. A., Young, M., and Douglas, T. (*2*005) Paramagnetic viral nanoparticles as potential high-relaxivity magnetic resonance contrast agents. *Magnetic Resonance in Medicine: An Official Journal of the International Society for Magnetic Resonance in Medicine*, **54**(4), 807-812.

5

Recent Advances in the Treatment of Infectious Diseases Using Nanoparticles

Santanu Sasidharan,[#] Shweta Raj[#] and Prakash Saudagar*

Department of Biotechnology, National Institute of Technology Warangal, Warangal, India

Corresponding author: ps@nitw.ac.in
[#]Authors contributed equally

5.1 Introduction

Nanotechnology is a nanoscale technology which involves the manipulation of matter at the atomic or molecular level. A nanoparticle is a single small object that behaves as a whole unit with respect to its transport and properties. Nanoscience is a multidisciplinary science of extremely small things at a nanoscale level which has a wide range of applications in areas like biology, medicine, chemistry, electronics, optics, physics, engineering and materials science.

A nanoparticle has size within 1-100 nm range with an interfacial surrounding layer. This interfacial layer is the outer covering and is generally responsible for the integral properties possessed by a nanoparticle. The concept of nanotechnology was seeded by the famous physicist Richard Feynman on December 29, 1959, in the American Physical Society meeting. His talk was entitled "There's Plenty of Room at the Bottom", which described a possible way for the synthesis of new materials by manipulating the matter at atomic or molecular levels. Officially, "nanotechnology" term was first used in 1974 by Norio Taniguchi. In the modern era of technology, nanotechnology started to grow significantly in 1980s after two major breakthrough developments in the field of nanoscience. The first breakthrough was the invention of a scanning tunneling microscope in 1981 by Gerd Binnig and Heinrich Rohrer, which allows the visualization of individual atoms and their bonding patterns (and later used for the atomic

Recent Trends in Nanobiotechnology, edited by Prakash Saudagar and K. Divakar
© 2019 Central West Publishing, Australia

manipulations). The second breakthrough was the discovery of full-erenes by Harry Kroto, Robert Curl and Richard Smalley in 1985. Full-erenes or carbon nanotubes are the carbon allotropes connected by covalent bonds in a hollow or spheroidal shape which have many potential applications in electrical and electronic devices. By mid-2000s, industrialization and commercialization of materials at nanoscale began to flourish at significant levels.

According to the FDA guidelines, a "nanomaterial" can be any product which shows the properties similar to a nanoparticle despite its size. Thus, even though the size of a material is more than 100 nm, these can be used for diagnostic purposes as nanomedicines. According to the European Science Foundation, nanomedicines are defined as "the science and technology of diagnosing, treating and preventing disease and traumatic injury, of relieving pain and of preserving and improving human health, using molecular tools and molecular knowledge of the human body". Later, the same definition was revised as "nanomedicine refers to highly specific medical intervention at the molecular scale for curing diseases or repairing damaged tissues, such as bone, muscle, or nerve" by National Institute of Health. Nanotechnology is extensively used in the medicinal field for medical devices and drug delivery systems. Different types of nanoparticles are used as nanorobots, nanobiosensors, therapeutic agents, dietary supplements and imaging systems for diagnostic purposes. In the last few decades, the global market of nanoparticles has boomed, and large sums have been invested in the field of life sciences alone. More precisely, nanoparticles assisted drug delivery approaches account for around 75% of nanotechnology investments and research. Nanoparticle formulations have many advantages over conventional medicinal approaches like improved efficacy and minimal side effects. Nanoparticles have specific targets depending on their atomic composition, physical properties and chemical structures, which make them a better approach for drug delivery actions. The first nanoparticle formulation-based drug delivery system was developed in 1978 as "liposome", and a large number of nanomaterial formulations have been developed since then for drug delivery and diagnostic purposes. Thus, various nanoparticles formulations are being used in the treatment of diseases like cancer, cardiovascular disorders, Alzheimer's and Parkinson's. During the course of evolution, microbes like bacteria, protozoans, fungi, parasites and viruses have developed resistance against the conventional drugs which necessitated the use of nanotechnology to develop novel anti-microbial drugs. Nanoparticle-based drugs are in great demand because of their specific drug targets, uniform bio-distribution and rapid onset of therapeutic action, which is mainly due to the sustained release of drugs depending on the intracellular conditions. These nanoparticle formulations avoid undesirable interactions and early degradation of drugs before reaching the target tissues. The other important advantage of nanoparticle-based drugs is that a single formulation can be used to deliver multi-drugs at the same time which can be used for the diseases with co-infections, like HIV with *Leishmania* or HIV with *Helicobacter pylori*. The major advantages of nanoparticle-based drugs are:

- high solubility
- high stability
- high surface area to volume ratio
- low microbial resistance
- less variability from patient to patient
- small dosage requirement
- non-toxicity

Therefore, nanoparticles offer an alternative route over conventional materials to overcome many challenges. The word "nano" is derived from the Latin word "nanus", which basically means dwarf. One nanometer is a unit of length which is equal to the one billionth of a meter (1 nm=10^{-9} meter). Typically, the bonds between two carbon atoms or interactions between any atoms or molecules are in the 0.12-0.15 nm range, whereas the diameter of a double helix DNA is around 2 nm and the length of Mycoplasma, the smallest cellular form of life, is around 200 nm. A representation of the nanoparticles with respect to size is given in Figure 5.1.

5.1.1 Different Types of Nanoparticles

Nanoparticles can be broadly classified based on their size, atomic compositions, morphology, physical and chemical properties. There are many techniques which are used for the fabrication of nanoparticles such as lithography, precipitation, degradation or agglomeration. Some of the nanoparticle platforms are carbon-based nanoparticles, metal nanoparticles, semiconductor nanoparticles, ceramic nanoparticles, lipid-based nanoparticles and polymeric nanoparticles. Nanoparticles are widely classified in three classes based on their dimensional arrangements at the atomic level.

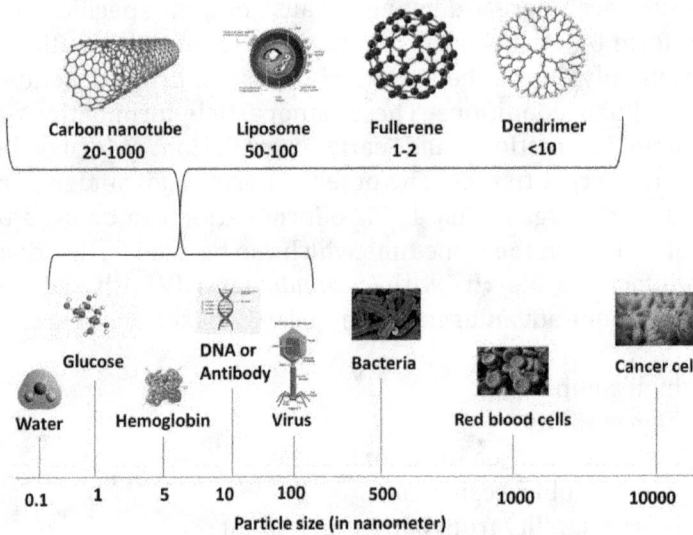

Figure 5.1 The classification of different nanoparticles based on their size. Nanoparticles in the region of 0-1000 nm in size are presented, along with other reference sizes for comparison.

1-D Nanoparticles

This class of nanoparticles, with the simplest arrangement of atoms, is generally arranged in the form of thin films or monolayers. These nanoparticles are widely used in solar cells, biosensors, storage devices and optical devices.

2-D Nanoparticles

Nanoparticles belonging to this class are often arranged in a cylindrical form or multilayered form. These are extensively used for thermal conductivity enhancement, peptide delivery and gene delivery.

3-D Nanoparticles

This type of nanoparticles is highly complex in nature and are generally represented as branched structures. In these branched networks, atoms are crosslinked with other nearby atoms or groups. These are mainly used in drug delivery and imaging techniques. A few nanoparticles and their applications are described in Table 5.1.

Table 5.1 The application of various nanoparticles and their characteristics. The dimensions of the nanoparticles define the characteristics of the nanoparticles.

Sr. No	Nano-particle type	Dimensions (nm)	Characteristics	Applications	Ref.
1	Carbon-based nanoparticles	Diameter: 0.5-3 Length: 20-1000	These include mainly two types of nanomaterials: carbon nanotubes (CNTs) and fullerenes. CNTs are the crystalline form of carbon sheets, either single layered (single-walled nanotubes, SWNTs) or multi-layered (multi-walled carbon nanotubes, MWCNTs). These particles show remarkable properties like high electrical conductivity, strength and electron affinity.	Cell-cell interactions, gene delivery and peptide delivery.	[1-3]
2	Metallic nanoparticles	<100	These are prepared by using their metal precursors. They can be easily synthesized by chemical or biological methods. These particles have very high surface energy because of the high surface area and can absorb small molecules.	Drug and gene delivery, biomolecular analysis, highly sensitive diagnostic assays, detection and imaging of biomolecules, etc.	[4,5]
3	Quantum dots	02-10	These are prepared using semiconductor materials and are used to improve optical properties and photostability. These consist of an inorganic semiconductor core with an outer coating of an aqueous shell. The inner core emits a bright fluorescence.	In-vitro live imaging of the organs, labeling of cancer markers, immunoassays, biosensors for diagnostic and DNA hybridization.	[6,7]
4	Liposomes	50-100	These are phospholipid-based vesicles that are	Cancer therapies, anti-leish-	[8,9]

			most developed and bi-ocompatible drug deliv-ery carriers. These lipid vesicles are synthesized by hydrating the dry phospholipids particles. These possess high en-trapment efficiency and flexibility.	maniasis, anti-malarial, deliv-ery of gene, peptide and protein, etc.	
5	Den-drimers	<10	These are highly branched monolayer polymers. These mainly consist of three parts, inner core, several mod-ified branches and outer surface coating. These are synthesized using different approaches of controlled polymeriza-tion.	Drug delivery targets, liver and pancreatic targeting, deliv-ery of proteins or peptides at cellular levels, etc.	[10,11]
6	Nano-poly-mers	10-1000	These are made up of several units linked with each other to form a polymeric unit. These are biodegradable and biocompatible in na-ture.	Drug coatings, controlled drug deliveries, ac-tive and pas-sive deliveries of bioactive particles at the target site, etc.	[4,12]

5.2 Characteristics of Nanoparticles

Nanoparticles are mainly designed on the basis of their size variation, morphology and surface charges. Different properties affect the phys-ical and chemical stability of the nanoparticles. Properties such as size distribution, particle diameter and surface charge affect the physical stability of the particles and are characterized by different microscopy techniques. Overall, different techniques have been used for the characterization and analysis of the physico-chemical proper-ties of the nanoparticles. These techniques include various types of microscopy methods like scanning electron microscopy (SEM), trans-mission electron microscopy (TEM), atomic force microscopy (AFM), X-ray photoelectron spectroscopy (XPS), X-ray diffraction (XRD) and zeta potential. In summary, the choice of the characterization tech-niques is mainly based on the physical and chemical properties shown by the nanoparticles.

5.2.1 Morphological Characterization

The properties of nanoparticles are mainly affected by morphological features. There are several techniques which are used for the morphological characterizations such as SEM and TEM. SEM is based on the principle of electron scanning and is widely used for the analysis of nanomaterial morphology. This technique is also used for the morphological analysis of the nanoparticles present in the matrices. For instance, the particle dispersion of SWNTs in polybutylene terephthalate and nylon-6 was studied using this technique [1]. Also, the analysis of ZnO modified metal-organic frameworks was performed using this technique, indicating the various morphologies of modified metal-organic frameworks at different reaction conditions [2]. TEM is based on the transmittance of electrons and provides the morphological analysis of the bulk materials even at low magnification. For instance, different morphologies of gold nanoparticles synthesized using different methods were studied using this technique [3,4].

5.2.2 Structural Characterization

This characterization includes the composition and bonding nature of the nanomaterials. This category includes techniques like XRD, zeta size analyzer, FT-IR, BET and Raman spectroscopy. One of the most important techniques for the structural characterization is XRD, which provides the crystal structure information. This technique has been widely used for the structural analysis of both single and multiphase nanoparticles [5]. On the other hand, XPS is one of the most sensitive techniques used for the determination of the exact elemental ratio as well as the bonding nature of the nanomaterials. The rotational and vibrational analysis of the bonds is usually carried out by FT-IR and Raman spectroscopy methods. For instance, FT-IR and XPS techniques were used for the analysis of alumina as a substrate in platinum nanoparticles [6].

5.2.3 Surface Area and Particle Size Characterization

Several techniques are used for the estimation of nanoparticle size which include SEM, TEM, AFM and XRD. Normally, microscopy techniques are used to determine the particle size, however, zeta analyzer is particularly used for the smallest size nanoparticles. The surface area of the nanoparticles is determined by the BET technique, which provides insights about the adsorption capacity of the particles. BET is based on the Brunauer-Emmett-Teller theorem, which works on the principle of adsorption and desorption. Generally, nitrogen gas is used for this the purpose to produce four specific isotherms, labeled as Type I, II, III and IV [7].

5.2.4 Optical Characterization

The optical properties possessed by any nanoparticle are of great interest in photo-catalytic applications. These properties are based on the light diffraction behavior of the materials. Optical characterization provides insights about light absorption, reflection, fluorescence and luminescence of the nanoparticles. These optical characterization techniques are especially useful for semiconductor nanoparticles and dendrimers. In one such study, optical analysis was carried out to study the electromagnetic absorption patterns of $LaFeO_3$ and La-FeO_3/MMT nanoparticles [8].

5.3 Classification of Nanoparticles

Nanoparticles are mainly classified as organic and inorganic nanoparticles. The classification is based on the type of material used for the synthesis process. Usually, carbon nanoparticles fall under the category of organic nanoparticles and magnetic/semiconductor nanoparticles are classified as inorganic nanoparticles. The mechanism of nanoparticles synthesis should be understood to control the size and size distribution, and it usually differs between different methods.

5.3.1 Inorganic Nanoparticles

The activity of the inorganic nanoparticles in biotechnology relies on their uniformity and biocompatibility with functional properties that support the ultimate usage. The inorganic nanoparticles are advantageous as compared to the organic nanoparticles because of their chemical and mechanical stability. Inorganic nanoparticles, like magnetic nanoparticles, can play effective role in cancer treatment, real-time visualization of biological events like cell signaling, interaction studies, etc. The synthesis of these nanoparticles is based on the two common approaches of nanoparticle synthesis, i.e., "top-down" approach involving attrition or milling and "bottom-up" approach that is basically assembles single atoms and molecules to resemble nanostructures. Top-down approach has heterogeneously sized products and impurities that make it a less attractive method of synthesis. The approach also results in crystallographic damages that eventually make it a non-viable option for various biological purposes. On the other hand, the bottom-up approach has higher chances of producing nanoparticles of homogenous size, thus, the probability of crystallographic defects is also decreased. The two major mechanisms that control the bottom-up approach method of synthesis are nucleation and growth mechanism. These two mechanisms have to be fine-tuned to control the size, size distribution and shape of the produced nanoparticles. The homogeneity plays an important role as far as the application in biopharmaceuticals is concerned, hence, the bottom-up approach of inorganic nanoparticles synthesis is preferred.

There are different methods for the synthesis of inorganic nanoparticles and each method is unique in terms of uniformity, size, yield and crystallinity of the particles. Few of the methods are listed below:

Salt Precipitation in an Aqueous Medium

The process basically involves the precipitation of metal salts in an aqueous medium by stoichiometry. The process has been commercially exploited for the synthesis of ferrous salt nanoparticles. The process parameters are still constantly studied and improved. Several other coating agents are also currently studied, such as citrate [9], polyethylene glycol (PEG) [10,11] and starch. Several other magnetic nanoparticles have also been synthesized using this approach, including cobalt ferrite [12], manganese ferrite [13], magnesium ferrite [14], cobalt oxide [15], manganese oxide [16] and nickel ferrite [17]. Further, polysaccharides including agarose, chitosan, heparin, starch, dextran and alginate have been used to coat the nanoparticles [18,19]. In another study, the stabilization of nanoparticles was achieved by co-precipitation at pH 7 using dimercaptosuccinic acid (DMSA) and silica [20]. The method has several advantages like ease of synthesis and cost effectiveness, along with control of particles size by stoichiometry, pH, stirring rate, temperature and ionic strength. It should be noted that the biopolymers used to coat the nanoparticles can strongly affect the behavior and can also change the toxicity profiles of the nanoparticles under study, thus, necessitating further exploration.

Hydrothermal Synthesis

The use of thermal and pressure properties of water to speed up precursor-based reaction is the basic mechanism underlying the hydrothermal synthesis. The method can be employed for both hydrophilic as well as hydrophobic molecules [21,22]. Moreover, the size and shape of the nanoparticles can be controlled. Hydrophilic nanoparticles can be synthesized by the use of citric acid. Ammonium iron citrate acts as a precursor in the presence of $N_2H_4.H_2O$. The process leads to the formation of isolated particles of approx. 4 nm size, whereas larger particles of 20 nm size are formed in the presence of urea. Urea, in this case, provides an alkaline environment for the nanoparticles [23]. Citric acid acts as a reducing agent of the metal ions, thus, the citric acid content determines the extent of reduction. Hydrothermal synthesis using organic precursors with fatty acids and a mixture of ethanol:water has also been shown to produce hydrophobic nanoparticles of various types [24,25]. The method, in general, requires synthesis at high temperatures.

Micro-emulsions

The stable isotropic dispersion of two immiscible liquids, wherein the stabilization is carried out by surfactant molecules, is achieved in this method. The water-oil emulsions can be reverse micelles which consist of micro-droplets of the metal salt solution in water. The stabilization is achieved by a surfactant, finally dispersed in a non-polar solvent. The surfactants used in this case are bis(2-ethylhexyl) sulphosuccinate (AOT), sodium dodecyl sulfate (SDS) and cetyltrimethylammonium bromide (CTAB). In a related study, solutions of Fe metal in ethanol were used for the synthesis of nanoparticles where SDS was used as surfactant and xylene was employed as oil phase [26]. Other metallic nanoparticles like Co and CoPt alloy have been prepared by using CTAB as surfactant and 1-butanol as co-surfactant, while octane is used as organic or oil phase [27]. A similar Fe synthesis method was reported using CTAB surfactant and octane organic phase in another study [28].

Polyol Process

The high-temperature reduction of metallic salts in the presence of a polyol, acting as solvent, surfactant and reducing agent, is carried out in this method. The physical properties of the nanoparticles are controlled by the kinetics of the precipitation. The polyalcohols used in this method include glycols and polyglycols. The polyols play the role of both solvent and reducing agent at the same time. These can also be used as stabilizing agent to prevents particle aggregation. The reaction is carried out in an alkaline environment, which helps the reduction process. The morphology of the nanoparticles can be controlled to a large extent. For instance, Fe nanoparticles with diverse shapes like cube, octahedron, sphere and equilateral octahedron have been synthesized by changing the concentration of the alkali [29]. In another study, $CoFe_2O_4$ nanoparticles with narrow particle size distribution have been obtained using mild conditions at 160 °C [30].

Decomposition in Organic Media

Thermal decomposition of magnetic nanoparticle salts in organic solvents with high boiling points and stabilizing surfactants produces nanoparticles with a high degree of crystallinity and narrow size distribution. The method involves the use of metal precursors with one or more stabilizers (carboxylates, amines and phosphine) and a reducing or oxidizing agent. The decomposition of the precursor takes place at elevated temperature, thus, the organic solvent needs to have a high boiling temperature. The hydrophobicity in the process is tackled by the addition of hydrophilic molecules or modification/substitution of surfactant coating. The process should be carried out in an inert atmosphere for achieving better control.

Aerosol Pyrolysis

In spray pyrolysis, the spraying of salt solution in the reactor is carried out, while undergoing evaporation of solvent. The process can also be performed using ultrasound. The aerosol droplet formation is carried out by nitrogen, ethylene and ammonia. The residence time of the micro-droplets in the reaction zone and concentration of the reaction is controlled by the gas flow into the reactor. The particles precipitated in this manner are further dried at high temperature. The final diameter of the particles is similar to the initial droplet size, usually 2-10 nm.

The laser pyrolysis method involves the heating of mixture prior to spraying to initiate nucleation. There are three characteristics in this method: small particle size, narrow size distribution and absence of aggregation. The nucleation process is fast, and the particles pushed into the reactor have less time to grow and crystallize. Therefore, the method is used to synthesize the nanoparticles of the very small size (5 nm) and homogeneous particles size distribution without any surfactants and additives.

Anisometric Particles

The process involves the synthesis of nanoparticles by controlling the morphology and properties. The diameter and length of the nanoparticles can be controlled by modifying the molar ratio of surfactants to metal precursors, the sequential addition of metal precursors and surfactants and other parameters like heating rate and reflux temperature. The three approaches generally used for the synthesis of the nanoparticles are:

- using hard templates such as silica, polymer spheres and metal oxides [31]
- using soft templates such as microemulsion droplets and gas bubbles [32]
- free templates such as Ostwald ripening mechanism, Kirkendall effect and self-attachment [33]

5.3.2 Organic Nanoparticles

Organic nanoparticles are solid nanoparticles (lipids or polymers) composed of organic compounds and have a size in the 10 nm to 1 μm range [34]. Compared to the inorganic nanoparticles, which are used as quantum dots, catalysts, etc., the organic nanoparticles have not received much attention due to different challenges. However, in recent years, the pharmaceutical industries have exploited the domain of organic nanoparticles for developing nano-medicines. This has led to the development of new materials and optimization of various techniques in this area. Similar to the inorganic nanoparticles, the synthesis of organic nanoparticles also involves two paths, i.e. "bottom-up" and "top-down" approaches. The bottom-up approach has been observed to possess encapsulation or active molecule carrier properties and has been used in the production of dendrimers, protein conjugates, DNA delivery machines, crosslinked polymer and liposomes. On the other hand, the top-down approach involves the grinding or milling down of the large particles to the required particle size distribution. Organic nanoparticles are ultimately soluble in water or other aqueous environments (if slowly), however, they are environment-friendly as compared to the inorganic nanoparticles. The synthesis of organic nanoparticles can be achieved using the following methods:

Emulsification Method

In this method, the mixture mainly contains two or more partially or fully immiscible liquids that may or may not employ an active surface agent for the synthesis. There are various kinds of mixture systems depending on the dispersion phase and medium, namely: oil in water (o/w), water in oil (w/o), oil in oil (o/o) or complex ones like w/o/w, o/w/o, w/o/o, etc. Depending on the droplet size of the emulsion, the emulsions can be classified as micro-emulsion (10-100 nm) and mini/macro-emulsion (100 nm-1 μm). Some of the methods commonly used to generate the nanoparticles using the emulsion methods are:

- solvent removal induced precipitation
- solvent evaporation method [35]
- solvent diffusion method [36]
- salting-out method [37]
- emulsion droplet generation [38]
- polymerization in emulsion [39]
- controlled radical polymerization [39]
- conventional emulsion polymerization [39]
- surfactant-free emulsion polymerization [39]
- interfacial polymerization

Nano-precipitation Method

The method was developed in the 1980s [40]. The technique is similar to the spontaneous emulsification technique, however, it is more eloquent, efficient and energy-efficient. The method involves the deposition of the polymer at the interface after the water soluble solvent is displaced from the lipophilic solution. There are three important constituents in this process: polymer, polymer solvent and polymer non-solvent. The polymer constituent can be synthetic, semi-synthetic or natural. The polymer solvent is selected depending upon two factors: high solubility in aqueous solution and effective evapo-ration. Several solvents like acetone, ethanol, hexane, methylene, di-oxane or a mixture of two or more solvents are employed [40-43]. The third phase is composed of a single non-solvent or a mixture of non-solvents, with an active surfactant. The mechanism of nanopar-ticle synthesis involves the diffusion of polymer solvent in the non-solvent phase by rapid mixing. The generated nanoparticles have properties dependent on the organic phase, injection rates, miscibil-ity of the organic phase, type and concentration of surface active agents and non-solvent phase. A wide range of polymers has been used for this method, such as polycaprolactone, polylactide, poly(hy-droxyl butyrate) and peptides [44-46].

Drying Method

The motivation to eliminate the use of organic solvents led to the de-velopment of spray drying process. The spray drying method in-volves two steps: rapid expansion of supercritical solution and rapid expansion of supercritical solution in the liquid solvent. This method has been used in recent years to generate micro-sized particles and convert nanoparticle suspensions into dry powders, which can even-tually be used in drug delivery [47,48]. The typical spray drying pro-cess consists of atomization of liquid into a spray of fine droplets that come into contact with hot air that dries the liquid to form sold parti-cles. The process has undergone significant evolution with time, and the synthesis of polymer nanoparticles in a one-step process has al-ready been successfully achieved. Preparation of polymer nanoparti-cles of Arabic gum, whey protein, malt-dextrin and poly(vinyl alco-hol) has also been achieved spray drying technology. The drying method leads to the development of organic nanoparticles which can be used in food and drug delivery applications. Several organic nano-particles have been deemed as "generally recognized as safe or GRAS" and few of these are:

- lipid-based GRAS [49]
- protein-based GRAS [50]
- polysaccharide-based GRAS [51]

The organic and inorganic nanoparticles produced by the de-scribed methods can eventually be used in pharmaceutical industries as drugs and drug carriers. The usage of these nanoparticles as drugs and drug delivery systems for the treatment of parasitic diseases is discussed below.

5.4 Role of Nanotechnology Against Infectious Diseases

The use of nanotechnology to cure infectious disease is gaining increasing importance. Many drugs have been discovered to combat diseases, however, these also lead to various side effects. In this respect, the nanoparticles have found applications as drugs, drug delivery systems and combinatorial drugs. The nanoparticle-based medicines exhibit the following mechanisms to eliminate the causative agents:

- binding to sulphur/phosphorus-containing biomolecules such as proteins and DNA
- mitochondrial deposition leading to oxidative stress pathway
- ion release from nanoparticles causing impairment of glycoprotein and lipophosphoglycan molecules by interaction with cysteine-containing proteins

Few of the diseases treated using nanoparticles as drugs are discussed below.

5.4.1 Leishmaniasis

Leishmaniasis is a protozoan disease that belongs to the *Leishmania* genus. The disease is transmitted by sandflies of *Phlebotomus* genus. There are three types of *Leishmania*, namely visceral leishmaniasis, cutaneous leishmaniasis and mucocutaneous leishmaniasis. Pentavalent antimonial and amphotericin are few of the drugs available for the treatment of the disease, however, these also cause significant side-effects. Silver has been used for long to treat various microbial and viral infections. Nano-silver, having the advantage of small size, high surface to volume ratio and penetration efficiency, has been utilized to treat cutaneous lesions [52,53]. Another strategy is to utilize amphotericin in its nano-form. The nanoparticles of amphotericin B have greater efficacy than conventional amphotericin B. The reduced costs, safety profile and increased efficacy may prove it to be an alternative to the conventional amphotericin B [54,55]. Nano-formulations of curcumin have been found to be effective against leishmaniasis independently as well as in combination with miltefosine in-vitro and in-vivo. The combination also exhibits an increase in lymphocyte proliferation [56]. Many metals and metal oxide nanoparticles have been observed to inhibit trypanothione metabolism enzymes that are vital for the parasite survival [57]. TiAg nanoparticles have also shown anti-leishmanial activity against *Leishmania tropica* and *Leishmania infantum*. The parasite is targeted by TiAg and visible light together to treat cutaneous leishmaniasis [58]. A combination of TiAg and *Nigella sativa* oil also displays high anti-leishmanial potency [59]. Nano-selenium has been reported to be anti-leishmanial against cutaneous leishmaniasis. The selenium nanoparticles inhibit the promastigote and amastigote form of *Leishmania* both in-vitro and in-vivo [60,61]. Several metallic salts have been used to treat leishmaniasis such as zinc sulfate, platinum (II), rhenium (V) and gold (III) [62,63]. Even though a variety of nanoparticles are available, the application of nanoparticles is, overall, still insufficient to treat leishmaniasis.

5.4.2 Malaria

Malaria is yet another serious disease caused by the parasite *Plasmodium*. The treatment of malarial diseases is limited due to drug resistance, parasite mutation, parasite overall load and co-infection by different strains of the parasite. Different metal nanoparticles have been employed for treatment due to their anti-malarial activity. For instance, silver nanoparticles can be used to treat malaria with significant activity against *Plasmodium falciparum* and vector Anopheles mosquito [64]. The inhibition of malaria has been achieved both in-vitro and in-vivo by silver nanoparticles [65-68]. In another study, gold nanoparticles have been reported to be potent against malaria *Plasmodium falciparum* [69,70]. The nanoparticles displayed an in-vivo delay in parasitemia and moderate anti-plasmodial activity. Other than gold and silver, metal oxide nanoparticles like Fe_3O_4, MgO, ZrO_2, Al_2O_3 and CeO_2 have been observed to exhibit moderate to high anti-plasmodial activity [71]. Numerous studies have also focused on restricting the growth of larvae and pupae of Anopheles by using silver nanoparticles [72-78].

5.4.3 Trypanosomiasis

Trypanosomes are the causative agents for the American and African forms of trypanosomiasis. African trypanosomiasis, caused by *Trypa-*

nosoma brucei and *Trypanosoma brucie gambiense*, is transmitted by the tsetse fly. American trypanosomiasis, also known as Chagas disease, is caused by *Trypanosoma cruzi*. The treatment is based on several drugs like suramin, pentamidine and melarsoprol, along with combinatorial treatment [79]. The limitations of such treatments include adverse effects, elevated toxicity and non-targeted mechanisms. Silver and gold nanoparticles have been assessed to target a critical enzyme like arginase kinase, leading to a decrement in activity by 7% [80,81]. Silver and gold nanoparticles also display inhibition effect on the growth of *T. brucei gambiense* [82]. Prodigiosin and violacein bacterial pigments have been coupled to Au and Ag nano-carriers and have exhibited anti-parasitic activity [83,84]. Prodigiosin shows higher activity that violacein when administered in combination with Au or Ag nanoparticles. Other than Au and Ag nanoparticles, only a few other nanomaterials have been observed to exhibit anti-trypanosomiasis activity.

5.5 Nanoparticles as Effective Drug Delivery Systems

Many nanosystems have been developed in the past few years in the therapeutic field. The systems work to deliver the drug dosage from the administration site to target site without any modification. The pharmaceutical industry has been trying to exploit this method of drug delivery. The nano-carriers are at the forefront of development in this field. Some of the nano-carrier systems to treat leishmaniasis, malaria and trypanosomiasis are at advanced stage of development. Nanomaterial assisted drug delivery of plant-based products is also gaining pace in the therapeutic industry to treat visceral leishmaniasis [85-87]. It has been found that artemisinin loaded nanoparticles possess improved anti-leishmanial activity ex-vivo [88]. Liposomal preparation has been shown as a better treatment against leishmaniasis when loaded with amphotericin B [89]. Another liposome-based treatment option with zinc phthalocyanine, a light sensitizer, has been shown to eliminate *Leishmania braziliensis* without affecting the macrophages [90]. Another study involving the preparation of nano-capsules by layer nano-emulsion method was observed to target *Leishmania donovani* when loaded with doxorubicin [91]. Poly(lactic-co-glycolic acid) (PLGA) nano-spheres containing amphotericin B also exhibit anti-leishmanial activity [92]. Poly(vinyl alcohol) (PVA) stabilized andrographolide, extracted from *Andrographis paniculata*, has been reported to exhibit high activity towards *Leishmania* cells

[93]. The drug is polymerized in PLGA at a drug:polymer ratio of 50:50.

In the treatment of malaria, egg phosphatidylcholine and cholesterol have been used as the first nano-carriers [94]. Phosphatidylglycerol (PG) and phosphatidylethanolamine (PE), conjugated with anti-malarial drugs, were also exploited recently. These liposomes were negatively charged and were found to be very efficient in targeting [95]. ART, an anti-malarial drug, was found to have increased efficiency and increased half-life when encapsulated in different phospholipids, mainly CHOL [96]. ART based nano-emulsions (ART-NE), prepared by high-pressure homogenization, enhanced its efficacy against malaria parasites. ART-NE were also found to be stable and had uniform size distribution [97]. Primaquine nano-emulsions, exhibited higher oral bio-availability, and the drug uptake was found to be 45% higher than the conventional drugs in the liver [98]. Clotrimazole, an anti-mycotic and anti-malarial drug, was injected into infected mice and exhibited higher onset of activity and percent reduction of parasitemia. This nano-emulsion can be used for the treatment of malaria in a dose-dependent manner [99]. Polymeric nanoparticles formed by chitosan loaded with dsRNA, with a particle size of 100-200 nm, showed 71% growth inhibition at the end of 48 h period, at a dose of 10 mg/ml [100]. AGMA1 and ISA23, which are poly(amidoamine) drug conjugates, exhibit a preferential binding and internalization in red blood cells (RBC's) [101]. In another study, monensin loaded nanoparticles exhibited a 10 times higher activity in inhibiting *Plasmodium falciparum*, when compared to free monensin [102]. Protein-protein conjugated nanoparticles for malarial antigen delivery were also studied recently and demonstrated better performance with respect to drug carrying function and vaccine development [103].

For trypanosomes or Chagas disease, a DNA vaccine has been developed, where immunogenicity is delivered by vaccine nanoparticle delivery system [104]. Another study based on pentamidine loaded chitosan nanoparticles reported the requirement of 100-fold lower curative dosage as compared to conventional pentamidine. The formulation displayed undiminished activity against mutated cell line resistant to pentamidine [105]. Diminazene, a trypanocidal drug, was precipitated by colloidal formulation in porous cationic nanoparticles with the oily core. The efficacy of colloidal diminazene was found to be increased, along with long term stability (six months) [106]. Another formulation of diminazene aceturate, functionalized with silver or gold nanoparticles, has also been reported to exhibit trypanocidal activity in a dose-dependent manner [107]. Arsonolipids containing a palmitic acid acyl chain were also evaluated for anti-trypanosoma activity [108-110]. Numerous other nano-carriers have also been formulated with different methods. However, the clinical trials and analysis of such nanoparticle loaded with drugs have to be carried out in detail for further development.

5.6 Current Clinical Trials and Market Analysis

The journey towards the development of the first nanoparticle-based drug delivery systems has been long. A few nano-drug delivery systems exist in the market today. The reasons for the slow growth of the drug delivery systems are:

- slow translation from labs to markets
- failure to compare the results of pre-clinical studies
- complex scaling up of lab scale experiments
- lack of cost-effective characterization of nanoparticles synthesized in each batch
- national and international regulations as well as lack of industrial support

In this respect, the large number of research studies in this field have not proportionally translated into clinical trials. A few drug delivery systems present at clinical trial levels today include ciprofloxacin liposomes, vaccine carriers for Ebola virus and silver nanoparticles in central venous catheters. Among the drugs and drug carriers that have passed the clinical trials to reach the commercial markets, amphotericin B liposome (Ambisome®), solid nanobase® to treat hepatitis C and virosomal vaccines like Inflexal V® and Epaxal® are the prominent examples. Several other nanoparticle-based diagnostic and medical devices, such as Verigene®, Silverline® and Endorem®, have also passed the trials

5.7 Conclusions and Future Perspective

Nanoparticles and nanoparticle-based drug delivery systems for the treatment of infectious diseases target the necessary site of action. The nanoparticle systems not only perform the carrier activity effectively, but also significantly enhance the effect of drugs, sometimes in drug-resistant infectious diseases. The nanoparticles hold several other advantages like lower dosage requirements, increased bio-availability and less adverse side-effects. Even though nanoparticles hold immense potential, one should also be mindful about their draw-backs such as scarce knowledge with respect to their effect on metabolism, toxicity and clearance. These drawbacks compounded with the difficulty in reproducing the same effects in clinical studies have slowed the successful development of nanoparticles as drugs and drug delivery systems.

The future research on the development of nanoparticles for therapeutic industry should be of interdisciplinary nature. An amalgam of physicists and health-care researchers like biologists and pharmacists can bring better therapeutic agents in the market. The laboratory results should accurately translate into industrial scale processes. The clinical trials by translational researchers and industrial support are crucial for this journey. On a whole, the research should be focused on re-designing the therapeutic industry through patient-based dose regime as well as enhancing the output of the functional nanoparticles.

References

1. Khan, I., Saeed, K., and Khan, I. (2017) Nanoparticles: Properties, applications and toxicities. *Arabian Journal of Chemistry*, doi: 10.1016/j.arabjc.2017.05.011.
2. Mirzadeh, E., and Akhbari, K. (2016) Synthesis of nanomaterials with desirable morphologies from metal-organic frameworks for various applications. *CrystEngComm*, **18**(39), 7410-7424.
3. Khlebtsov, N. G., and Dykman, L. A. (2010) Optical properties and biomedical applications of plasmonic nanoparticles. *Journal of Quantitative Spectroscopy and Radiative Transfer*, **111**(1), 1-35.
4. Khlebtsov, N., and Dykman, L. (2011) Biodistribution and toxicity of engineered gold nanoparticles: a review of in vitro and in vivo studies. *Chemical Society Reviews*, **40**(3), 1647-1671.
5. Emery, A. A., Saal, J. E., Kirklin, S., Hegde, V. I., and Wolverton, C. (2016) High-throughput computational screening of perovskites for thermochemical water aplitting applications. *Chemistry of Materials*, **28**(16), 5621-5634.
6. Dablemont, C., Lang, P., Mangeney, C., Piquemal, J.-Y., Petkov, V., Herbst, F., and Viau, G. (2008) FTIR and XPS study of Pt nanoparticle functionalization and interaction with alumina. *Langmuir*, **24**(11), 5832-5841.
7. Zhou, M., Wei, Z., Qiao, H., Zhu, L., Yang, H., and Xia, T. (2009) Particle

size and pore structure characterization of silver nanoparticles prepared by confined arc plasma. *Journal of Nanomaterials*, **2009**, Article ID 968058.

8. Peng, K., Fu, L., Yang, H., and Ouyang, J. (2016) Perovskite La-FeO$_3$/montmorillonite nanocomposites: synthesis, interface characteristics and enhanced photocatalytic activity. *Scientific Reports*, **6**, 19723.

9. Lesieur, S., Grabielle-Madelmont, C., Ménager, C., Cabuil, V., Dadhi, D., Pierrot, P., and Edwards, K. (2003) Evidence of surfactant-induced formation of transient pores in lipid bilayers by using magnetic-fluid-loaded liposomes. *Journal of the American Chemical Society*, **125**(18), 5266-5267.

10. Kim, D. K., Mikhaylova, M., Zhang, Y., and Muhammed, M. (2003) Protective coating of superparamagnetic iron oxide nanoparticles. *Chemistry of Materials*, **15**(8), 1617-1627.

11. Liu, X., Guan, Y., Ma, Z., and Liu, H. (2004) Surface modification and characterization of magnetic polymer nanospheres prepared by miniemulsion polymerization. *Langmuir*, **20**(23), 10278-10282.

12. Veverka, M., Veverka, P., Kaman, O., Lančok, A., Závěta, K., Pollert, E., Knížek, K., Boháček, J., Beneš, M., and Kašpar, P. (2007) Magnetic heating by cobalt ferrite nanoparticles. *Nanotechnology*, **18**(34), 345704.

13. Tourinho, F. A., Franck, R., and Massart, R. (1990) Aqueous ferrofluids based on manganese and cobalt ferrites. *Journal of Materials Science*, **25**(7), 3249-3254.

14. Chen, Q., Rondinone, A. J., Chakoumakos, B. C., and Zhang, Z. J. (1999) Synthesis of superparamagnetic MgFe2O4 nanoparticles by coprecipitation. *Journal of Magnetism and Magnetic Materials*, **194**(1-3), 1-7.

15. Sinkó, K., Szabó, G. and Zrínyi, M. (2011) Liquid-phase synthesis of cobalt oxide nanoparticles. *Journal of Nanoscience and Nanotechnology*, **11**(5), 4127-4135.

16. Portehault, D., Cassaignon, S., Baudrin, E., and Jolivet, J.-P. (2009) Structural and morphological control of manganese oxide nanoparticles upon soft aqueous precipitation through MnO$_4^-$/Mn^{2+} reaction. *Journal of Materials Chemistry*, **19**(16), 2407-2416.

17. Regazzoni, A. E., and Matijević, E. (1982) Formation of spherical colloidal nickel ferrite particles as model corrosion products. *Corrosion*, **38**(4), 212-218.

18. Dias, A., Hussain, A., Marcos, A., and Roque, A. (2011) A biotechnological perspective on the application of iron oxide magnetic colloids modified with polysaccharides. *Biotechnology Advances*, **29**(1), 142-155.

19. De la Fuente, J. M., Alcantara, D., and Penades, S. (2007) Cell response to magnetic glyconanoparticles: does the carbohydrate mat-

ter? *IEEE Transactions On Nanobioscience*, **6**(4), 275-281.

20. Fauconnier, N., Pons, J., Roger, J., and Bee, A. (1997) Thiolation of maghemite nanoparticles by dimercaptosuccinic acid. *Journal of Colloid and Interface Science*, **194**(2), 427-433.

21. Yan, J., Mo, S., Nie, J., Chen, W., Shen, X., Hu, J., Hao, G., and Tong, H. (2009) Hydrothermal synthesis of monodisperse Fe_3O_4 nanoparticles based on modulation of tartaric acid. *Colloids and Surfaces A: Physicochemical and Engineering Aspects*, **340**(1-3), 109-114.

22. Wang, X., Zhuang, J., Peng, Q., and Li, Y. (2005) A general strategy for nanocrystal synthesis. *Nature*, **437**(7055), 121-124.

23. Cheng, W., Tang, K., Qi, Y., Sheng, J., and Liu, Z. (2010) One-step synthesis of superparamagnetic monodisperse porous Fe_3O_4 hollow and core-shell spheres. *Journal of Materials Chemistry*, **20**(9), 1799-1805.

24. Ge, J.-P., Xu, S., Zhuang, J., Wang, X., Peng, Q., and Li, Y.-D. (2006) Synthesis of CdSe, ZnSe, and $Zn_xCd_{1-x}Se$ nanocrystals and their silica sheathed core/shell structures. *Inorganic Chemistry*, **45**(13), 4922-4927.

25. Liang, X., Wang, X., Zhuang, J., Chen, Y., Wang, D., and Li, Y. (2006) Synthesis of nearly monodisperse iron oxide and oxyhydroxide nanocrystals. *Advanced Functional Materials*, **16**(14), 1805-1813.

26. Lee, Y., Lee, J., Bae, C. J., Park, J. G., Noh, H. J., Park, J. H., and Hyeon, T. (2005) Large scale synthesis of uniform and crystalline magnetite nanoparticles using reverse micelles as nanoreactors under reflux conditions. *Advanced Functional Materials*, **15**(3), 503-509.

27. Carpenter, E. E., Seip, C. T., and O'Connor, C. J. (1999) Magnetism of nanophase metal and metal alloy particles formed in ordered phases. *Journal of Applied Physics*, **85**(8), 5184-5186.

28. Carpenter, E. E. (2001) Iron nanoparticles as potential magnetic carriers. *Journal of Magnetism and Magnetic Materials*, **225**(1-2), 17-20.

29. Zhao, L., Zhang, H., Xing, Y., Song, S., Yu, S., Shi, W., Guo, X., Yang, J., Lei, Y., and Cao, F. (2007) Morphology-controlled synthesis of magnetites with nanoporous structures and excellent magnetic properties. *Chemistry of Materials*, **20**(1), 198-204.

30. Ammar, S., Helfen, A., Jouini, N., Fievet, F., Rosenman, I., Villain, F., Molinie, P., and Danot, M. (2001) Magnetic properties of ultrafine cobalt ferrite particles synthesized by hydrolysis in a polyol medium. *Journal of Materials Chemistry*, **11**(1), 186-192.

31. Kim, S.-W., Kim, M., Lee, W.Y., and Hyeon, T. (2002) Fabrication of hollow palladium spheres and their successful application to the recyclable heterogeneous catalyst for Suzuki coupling reactions. *Journal of the American Chemical Society*, **124**(26), 7642-7643.

32. Zhang, X., and Li, D. (2006) Metal compound induced vesicles as efficient directors for rapid synthesis of hollow alloy spheres. *Angew-*

andte Chemie International Edition, **45**(36), 5971-5974.
33. Yin, Y., Rioux, R. M. Erdonmez, C. K. Hughes, S. Somorjai, G. A. and Alivisatos, A. P. (2004) Formation of hollow nanocrystals through the nanoscale Kirkendall effect. *Science*, **304**(5671), 711-714.
34. Drexler, K. E. (1981) Molecular engineering: An approach to the development of general capabilities for molecular manipulation. *Proceedings of the National Academy of Sciences*, **78**(9), 5275-5278.
35. Gurny, R., Peppas, N., Harrington, D., and Banker, G. (1981) Development of biodegradable and injectable latices for controlled release of potent drugs. *Drug Development and Industrial Pharmacy*, **7**(1), 1-25.
36. Quintanar-Guerrero, D., Allémann, E. Doelker, E. and Fessi, H. (1998) Preparation and characterization of nanocapsules from preformed polymers by a new process based on emulsification-diffusion technique. *Pharmaceutical Research*, **15**(7), 1056-1062.
37. Ibrahim, H., Bindschaedler, C. Doelker, E. Buri, P. and Gurny, R. (1992) Aqueous nanodispersions prepared by a salting-out process. *International Journal of Pharmaceutics*, **87**(1-3), 239-246.
38. Müller, R. H., MaÈder, K. and Gohla, S. (2000) Solid lipid nanoparticles (SLN) for controlled drug delivery-a review of the state of the art. *European Journal of Pharmaceutics and Biopharmaceutics*, **50**(1), 161-177.
39. Landfester, K. (2009) Miniemulsion polymerization and the structure of polymer and hybrid nanoparticles. *Angewandte Chemie International Edition*, **48**(25), 4488-4507.
40. Fessi, H., Puisieux, F. Devissaguet, J.P. Ammoury, N. and Benita, S. (1989) Nanocapsule formation by interfacial polymer deposition following solvent displacement. *International Journal of Pharmaceutics*, **55**(1), R1-R4.
41. Chang, J., Jallouli, Y. Kroubi, M. Yuan, X.-b. Feng, W. Kang, C.-s. Pu, P.-y. and Betbeder, D. (2009) Characterization of endocytosis of transferrin-coated PLGA nanoparticles by the blood-brain barrier. *International Journal of Pharmaceutics*, **379**(2), 285-292.
42. Nassar, T., Rom, A., Nyska, A., and Benita, S. (2009) Novel double coated nanocapsules for intestinal delivery and enhanced oral bioavailability of tacrolimus, a P-gp substrate drug. *Journal of Controlled Release*, **133**(1), 77-84.
43. De Assis, D. N., Mosqueira, V. C. F., Vilela, J. M. C., Andrade, M. S., and Cardoso, V. N. (2008) Release profiles and morphological characterization by atomic force microscopy and photon correlation spectroscopy of 99mTechnetium-fluconazole nanocapsules. *International Journal of Pharmaceutics*, **349**(1-2), 152-160.
44. Moinard-Chécot, D., Chevalier, Y., Briançon, S., Beney, L., and Fessi, H. (2008) Mechanism of nanocapsules formation by the emulsion-diffusion process. *Journal of Colloid and Interface Science*, **317**(2),

458-468.
45. Zili, Z., Sfar, S., and Fessi, H. (2005) Preparation and characterization of poly-ε-caprolactone nanoparticles containing griseofulvin. *International Journal of Pharmaceutics*, **294**(1-2), 261-267.
46. Yallapu, M. M., Gupta, B. K., Jaggi, M., and Chauhan, S. C. (2010) Fabrication of curcumin encapsulated PLGA nanoparticles for improved therapeutic effects in metastatic cancer cells. *Journal of Colloid and Interface Science*, **351**(1), 19-29.
47. Li, X., Anton, N., Arpagaus, C., Belleteix, F., and Vandamme, T. F. (2010) Nanoparticles by spray drying using innovative new technology: The Büchi nano spray dryer B-90. *Journal of Controlled Release*, **147**(2), 304-310.
48. Lee, S. H., Heng, D., Ng, W. K., Chan, H.-K., and Tan, R. B. (2011) Nano spray drying: a novel method for preparing protein nanoparticles for protein therapy. *International Journal Of Pharmaceutics*, **403**(1-2), 192-200.
49. Tallury, P., Malhotra, A., Byrne, L. M., and Santra, S. (2010) Nanobioimaging and sensing of infectious diseases. *Advanced Drug Delivery Reviews*, **62**(4-5), 424-437.
50. Bittner, A. M. (2005) Biomolecular rods and tubes in nanotechnology. *Naturwissenschaften*, **92**(2), 51-64.
51. Chayed, S., and Winnik, F. M. (2007) In vitro evaluation of the mucoadhesive properties of polysaccharide-based nanoparticulate oral drug delivery systems. *European Journal of Pharmaceutics and Biopharmaceutics*, **65**(3), 363-370.
52. Santos, D. M., Carneiro, M. W., de Moura, T. R., Fukutani, K., Clarencio, J., Soto, M., Espuelas, S., Brodskyn, C., Barral, A., and Barral-Netto, M. (2012) Towards development of novel immunization strategies against leishmaniasis using PLGA nanoparticles loaded with kinetoplastid membrane protein-11. *International Journal of Nanomedicine*, **7**, 2115.
53. Kunjachan, S., Jose, S., Thomas, C. A., Joseph, E., Kiessling, F., and Lammers, T. (2012) Physicochemical and biological aspects of macrophage mediated drug targeting in anti-microbial therapy. *Fundamental and Clinical Pharmacology*, **26**(1), 63-71.
54. Manandhar, K. D., Yadav, T. P., Prajapati, V. K., Kumar, S., Rai, M., Dube, A., Srivastava, O. N., and Sundar, S. (2008) Antileishmanial activity of nano-amphotericin B deoxycholate. *Journal of Antimicrobial Chemotherapy*, **62**(2), 376-380.
55. Shahcheraghi, S., Ayatollahi, J., Lotfi, M., Bafghi, A., and Khaleghinejad, S. (2016) Application of nano drugs in treatment of leishmaniasis. *Global Journal of Infectious Diseases and Clinical Research*, **2**(1), 018-020.
56. Tiwari, B., Pahuja, R., Kumar, P., Rath, S. K., Gupta, K. C., and Goyal, N. (2017) Nanotized curcumin and miltefosine, a potential combin-

ation for treatment of experimental visceral leishmaniasis. *Antimicrobial Agents and Chemotherapy*, **61**(3), e01169-16.

57. Navarro, M., Gabbiani, C., Messori, L., and Gambino, D. (2010) Metal-based drugs for malaria, trypanosomiasis and leishmaniasis: recent achievements and perspectives. *Drug Discovery Today*, **15**(23-24), 1070-1078.

58. Allahverdiyev, A. M., Abamor, E. S., Bagirova, M., Baydar, S. Y., Ates, S. C., Kaya, F., Kaya, C., and Rafailovich, M. (2013) Investigation of antileishmanial activities of TiO2@Ag nanoparticles on biological properties of L. tropica and L. infantum parasites, in vitro. *Experimental Parasitology*, **135**(1), 55-63.

59. Abamor, E. S., and Allahverdiyev, A. M. (2016) A nanotechnology based new approach for chemotherapy of cutaneous leishmaniasis: TIO2@Ag nanoparticles-Nigella sativa oil combinations. *Experimental Parasitology*, **166**, 150-163.

60. Beheshti, N., Soflaei, S., Shakibaie, M., Yazdi, M. H., Ghaffarifar, F., Dalimi, A., and Shahverdi, A. R. (2013) Efficacy of biogenic selenium nanoparticles against Leishmania major: in vitro and in vivo studies. *Journal of Trace Elements in Medicine and Biology*, **27**(3), 203-207.

61. Mahmoudvand, H., Shakibaie, M., Tavakoli, R., Jahanbakhsh, S., and Sharifi, I. (2014) In vitro study of leishmanicidal activity of biogenic selenium nanoparticles against Iranian isolate of sensitive and glucantime-resistant Leishmania tropica. *Iranian Journal of Parasitology*, **9**(4), 452.

62. Minodier, P., and Parola, P. (2007) Cutaneous leishmaniasis treatment. *Travel Medicine and Infectious Disease*, **5**(3), 150-158.

63. Fricker, S. P., Mosi, R. M., Cameron, B. R., Baird, I., Zhu, Y., Anastassov, V., Cox, J., Doyle, P. S., Hansell, E., and Lau, G. (2008) Metal compounds for the treatment of parasitic diseases. *Journal of Inorganic Biochemistry*, **102**(10), 1839-1845.

64. Rai, M., Ingle, A.P., Paralikar, P., Gupta, I., Medici, S., and Santos, C. A. (2017) Recent advances in use of silver nanoparticles as antimalarial agents. *International Journal of Pharmaceutics*, **526**(1-2), 254-270.

65. Mishra, A., Kaushik, N. K., Sardar, M., and Sahal, D. (2013) Evaluation of antiplasmodial activity of green synthesized silver nanoparticles. *Colloids and Surfaces B: Biointerfaces*, **111**, 713-718.

66. Jaganathan, A., Murugan, K., Panneerselvam, C., Madhiyazhagan, P., Dinesh, D., Vadivalagan, C., Chandramohan, B., Suresh, U., Rajaganesh, R., and Subramaniam, J. (2016) Earthworm-mediated synthesis of silver nanoparticles: A potent tool against hepatocellular carcinoma, Plasmodium falciparum parasites and malaria mosquitoes. *Parasitology International*, **65**(3), 276-284.

67. Murugan, K., Panneerselvam, C., Subramaniam, J., Madhiyazhagan,

P., Hwang, J.-S., Wang, L., Dinesh, D., Suresh, U., Roni, M., and Higuchi, A. (2016) Eco-friendly drugs from the marine environment: sponge-weed-synthesized silver nanoparticles are highly effective on Plasmodium falciparum and its vector Anopheles stephensi, with little non-target effects on predatory copepods. *Environmental Science and Pollution Research*, **23**(16), 16671-16685.

68. Murugan, K., Panneerselvam, C., Samidoss, C. M., Madhiyazhagan, P., Suresh, U., Roni, M., Chandramohan, B., Subramaniam, J., Dinesh, D., and Rajaganesh, R. (2016) In vivo and in vitro effectiveness of Azadirachta indica-synthesized silver nanocrystals against Plasmodium berghei and Plasmodium falciparum, and their potential against malaria mosquitoes. *Research In Veterinary Science*, **106**, 14-22.

69. Karthik, L., Kumar, G., Keswani, T., Bhattacharyya, A., Reddy, B. P., and Rao, K. B., (2013) Marine actinobacterial mediated gold nanoparticles synthesis and their antimalarial activity. *Nanomedicine: Nanotechnology, Biology and Medicine*, **9**(7), 951-960.

70. Dutta, P. P., Bordoloi, M., Gogoi, K., Roy, S., Narzary, B., Bhattacharyya, D. R., Mohapatra, P. K., and Mazumder, B. (2017) Antimalarial silver and gold nanoparticles: Green synthesis, characterization and in vitro study. *Biomedicine and Pharmacotherapy*, **91**, 567-580.

71. Inbaneson, S. J., and Ravikumar, S. (2013) In vitro antiplasmodial activity of PDDS-coated metal oxide nanoparticles against Plasmodium falciparum. *Applied Nanoscience*, **3**(3), 197-201.

72. Arokiyaraj, S., Kumar, V. D., Elakya, V., Kamala, T., Park, S. K., Ragam, M., Saravanan, M., Bououdina, M., Arasu, M. V., and Kovendan, K. (2015) Biosynthesized silver nanoparticles using floral extract of Chrysanthemum indicum L. - Potential for malaria vector control. *Environmental Science and Pollution Research*, **22**(13), 9759-9765.

73. Dinesh, D., Murugan, K., Madhiyazhagan, P., Panneerselvam, C., Kumar, P. M., Nicoletti, M., Jiang, W., Benelli, G., Chandramohan, B., and Suresh, U. (2015) Mosquitocidal and antibacterial activity of green-synthesized silver nanoparticles from Aloe vera extracts: towards an effective tool against the malaria vector Anopheles stephensi? *Parasitology Research*, **114**(4), 1519-1529.

74. Kumar, K. R., Nattuthurai, N., Gopinath, P., and Mariappan, T. (2015) Synthesis of eco-friendly silver nanoparticles from Morinda tinctoria leaf extract and its larvicidal activity against Culex quinquefasciatus. *Parasitology Research*, **114**(2), 411-417.

75. Nyakundi, E. O., and Padmanabhan, M. N. (2015) Green chemistry focus on optimization of silver nanoparticles using response surface methodology (RSM) and mosquitocidal activity: Anopheles stephensi (Diptera: Culicidae). *Spectrochimica Acta Part A, Molecular and Biomolecular Spectroscopy*, **149**, 978-984.

76. Poopathi, S., De Britto, L. J., Praba, V. L., Mani, C., and Praveen, M. (2015) Synthesis of silver nanoparticles from Azadirachta indica - A most effective method for mosquito control. *Environmental Science and Pollution Research*, **22**(4), 2956-2963.

77. Santhosh, S., Yuvarajan, R., and Natarajan, D. (2015) Annona muricata leaf extract-mediated silver nanoparticles synthesis and its larvicidal potential against dengue, malaria and filariasis vector. *Parasitology Research*, **114**(8), 3087-3096.

78. Soni, N., and Prakash, S. (2014) Silver nanoparticles: a possibility for malarial and filarial vector control technology. *Parasitology Research*, **113**(11), 4015-4022.

79. Murthy, S., Keystone, J., and Kissoon, N. (2013) Infections of the developing world. *Critical Care Clinics*, **29**(3), 485-507.

80. Miranda, M. R., Canepa, G. E., Bouvier, L. A., and Pereira, C. A. (2006) Trypanosoma cruzi: Oxidative stress induces arginine kinase expression. *Experimental Parasitology*, **114**(4), 341-344.

81. Adeyemi, O. S., and Whiteley, C. G. (2013) Interaction of nanoparticles with arginine kinase from Trypanosoma brucei: kinetic and mechanistic evaluation. *International Journal of Biological Macromolecules*, **62**, 450-456.

82. Rahul, S., Chandrashekhar, P., Hemant, B., Bipinchandra, S., Mouray, E., Grellier, P., and Satish, P. (2015) In vitro antiparasitic activity of microbial pigments and their combination with phytosynthesized metal nanoparticles. *Parasitology International*, **64**(5), 353-356.

83. Lopes, S. C., Blanco, Y. C., Justo, G. Z., Nogueira, P. A., Rodrigues, F. L., Goelnitz, U., Wunderlich, G., Facchini, G., Brocchi, M., and Duran, N. (2009) Violacein extracted from Chromobacterium violaceum inhibits Plasmodium growth in vitro and in vivo. *Antimicrobial Agents and Chemotherapy*, **53**(5), 2149-2152.

84. Durán, M., Faljoni-Alario, A., and Durán, N. (2010) Chromobacterium violaceum and its important metabolites. *Folia Microbiologica*, **55**(6), 535-547.

85. Nilforoushzadeh, M. A., Shirani-Bidabadi, L., Zolfaghari-Baghbaderani, A., Jafari, R., Heidari-Beni, M., Siadat, A. H., and Ghahraman-Tabrizi, M. (2012) Topical effectiveness of different concentrations of nanosilver solution on Leishmania major lesions in Balb/c mice. *Journal of Vector Borne Diseases*, **49**(4), 249.

86. Cardona-Arias, J. A., Vélez, I. D., and López-Carvajal, L. (2015) Efficacy of thermotherapy to treat cutaneous leishmaniasis: a meta-analysis of controlled clinical trials. *Plos One*, **10**(5), e0122569.

87. Matin, R. (2015) Cutaneous leishmaniasis-treatment options in children. *British Journal of Dermatology*, **172**(4), 844-845.

88. Want, M. Y., Islamuddin, M., Chouhan, G., Ozbak, H. A., Hemeg, H. A., Dasgupta, A. K., Chattopadhyay, A. P., and Afrin, F. (2015) Therapeutic efficacy of artemisinin-loaded nanoparticles in experimental vis-

ceral leishmaniasis. *Colloids and Surfaces B: Biointerfaces*, **130**, 215-221.

89. Vyas, S. P., and Gupta, S. (2006) Optimizing efficacy of amphotericin B through nanomodification. *International Journal of Nanomedicine*, **1**(4), 417.

90. Perez, A. P., Casasco, A., Schilrreff, P., Tesoriero, M. V. D., Duempel-mann, L., Altube, M. J., Higa, L., Morilla, M. J., Petray, P., and Romero, E. L. (2014) Enhanced photodynamic leishmanicidal activity of hydrophobic zinc phthalocyanine within archaeolipids containing liposomes. *International Journal of Nanomedicine*, **9**, 3335.

91. Kansal, S., Tandon, R., Verma, A., Misra, P., Choudhary, A., Verma, R., Verma, P., Dube, A., and Mishra, P. (2014) Coating doxorubicin loaded nanocapsules with alginate enhances therapeutic efficacy against L eishmania in hamsters by inducing Th1 type immune responses. *British Journal of Pharmacology*, **171**(17), 4038-4050.

92. Farokhzad, O. C., and Langer, R. (2009) Impact of nanotechnology on drug delivery. *ACS Nano*, **3**(1), 16-20.

93. Roy, P., Das, S., Bera, T., Mondol, S., and Mukherjee, A. (2010) Andrographolide nanoparticles in leishmaniasis: characterization and in vitro evaluations. *International Journal of Nanomedicine*, **5**, 1113-1121.

94. Pirson, P., Steiger, R., Trouet, A., Gillet, J., and Herman, F. (1979) Liposomes in the chemotherapy of experimental murine malaria. *Transactions of the Royal Society of Tropical Medicine and Hygiene*, **73**(3), 347.

95. Qiu, L., Jing, N., and Jin, Y. (2008) Preparation and in vitro evaluation of liposomal chloroquine diphosphate loaded by a transmembrane pH-gradient method. *International Journal of Pharmaceutics*, **361**(1-2), 56-63.

96. Al-Angary, A., Al-Meshal, M., Bayomi, M., and Khidr, S. (1996) Evaluation of liposomal formulations containing the antimalarial agent, arteether. *International Journal of Pharmaceutics*, **128**(1-2), 163-168.

97. Dwivedi, P., Khatik, R., Chaturvedi, P., Khandelwal, K., Taneja, I., Raju, K. S. R., Dwivedi, H., Singh, S. K., Gupta, P. K., and Shukla, P. (2015) Arteether nanoemulsion for enhanced efficacy against Plasmodium yoelii nigeriensis malaria: An approach by enhanced bioavailability. *Colloids and Surfaces B: Biointerfaces*, **126**, 467-475.

98. Singh, K. K., and Vingkar, S. K. (2008) Formulation, antimalarial activity and biodistribution of oral lipid nanoemulsion of primaquine. *International Journal of Pharmaceutics*, **347**(1-2), 136-143.

99. Borhade, V., Pathak, S., Sharma, S., and Patravale, V. (2012) Clotrimazole nanoemulsion for malaria chemotherapy. Part I: preformulation studies, formulation design and physicochemical evaluation. *International Journal of Pharmaceutics*, **431**(1-2), 138-148.

100. Attasart, P., Boonma, S., Sunintaboon, P., Tanwilai, D., Pothikasikorn, J., and Noonpakdee, W. T. (2016) Inhibition of Plasmodium falciparum proliferation in vitro by double-stranded RNA nanoparticle against malaria topoisomerase II. *Experimental Parasitology*, **164**, 84-90.

101. Urban, P., Valle-Delgado, J. J., Mauro, N., Marques, J., Manfredi, A., Rottmann, M., Ranucci, E., Ferruti, P., and Fernandez-Busquets, X. (2014) Use of poly(amidoamine) drug conjugates for the delivery of antimalarials to Plasmodium. *Journal of Controlled Release*, **177**, 84-95.

102. Surolia, R., Pachauri, M., and Ghosh, P. C. (2012) Preparation and characterization of monensin loaded PLGA nanoparticles: in vitro anti-malarial activity against Plasmodium falciparum. *Journal of Biomedical Nanotechnology*, **8**(1), 172-181.

103. Scaria, P. V., Chen, B., Rowe, C. G., Jones, D. S., Barnafo, E., Fischer, E. R., anderson, C., MacDonald, N. J., Lambert, L., Rausch, K. M., Narum, D. L., and Duffy, P. E. (2017) Protein-protein conjugate nanoparticles for malaria antigen delivery and enhanced immunogenicity. *PLoS One*, **12**(12), e0190312.

104. Barry, M. A., Wang, Q., Jones, K. M., Heffernan, M. J., Buhaya, M. H., Beaumier, C. M., Keegan, B. P., Zhan, B., Dumonteil, E., Bottazzi, M. E., and Hotez, P. J. (2016) A therapeutic nanoparticle vaccine against Trypanosoma cruzi in a BALB/c mouse model of Chagas disease. *Journalo of Human Vaccines and Immunotherapeutics*, **12**(4), 976-987.

105. Unciti-Broceta, J. D., Arias, J. L., Maceira, J., Soriano, M., Ortiz-Gonzalez, M., Hernandez-Quero, J., Munoz-Torres, M., de Koning, H. P., Magez, S., and Garcia-Salcedo, J. A. (2015) Specific cell targeting therapy bypasses drug resistance mechanisms in African trypanosomiasis. *PLoS Pathogens*, **11**(6), e1004942.

106. Kroubi, M., Daulouede, S. Karembe, H. Jallouli, Y. Howsam, M. Mossalayi, D. Vincendeau, P. and Betbeder, D. (2010) Development of a nanoparticulate formulation of diminazene to treat African trypanosomiasis. *Nanotechnology*, **21**(50), 505102.

107. Adeyemi, O. S., and Sulaiman, F. A. (2015) Evaluation of metal nanoparticles for drug delivery systems. *Journal of Biomedical Research*, **29**(2), 145-149.

108. Antimisiaris, S. G., Ioannou, P. V., and Loiseau, P. M. (2003) In-vitro antileishmanial and trypanocidal activities of arsonoliposomes and preliminary in-vivo distribution in BALB/c mice. *Journal of Pharmacy and Pharmacology*, **55**(5), 647-652.

109. Zagana, P., Klepetsanis, P., Ioannou, P. V., Loiseau, P. M., and Antimisiaris, S. G. (2007) Trypanocidal activity of arsonoliposomes: effect of vesicle lipid composition. *Biomedicine and Pharmacotherapy*, **61**(8), 499-504.

110. Zagana, P., Haikou, M., Klepetsanis, P., Giannopoulou, E., Ioannou, P. V., and Antimisiaris, S. G. (2008) In vivo distribution of arsonoliposomes: effect of vesicle lipid composition. *International Journal of Pharmaceutics*, **347**(1-2), 86-92.

Index

X

www.ingramcontent.com/pod-product-compliance
Lightning Source LLC
Chambersburg PA
CBHW050125240326
41458CB00122B/1401